International Conference on Mechanical Engineering and Modern Technologies (MEMT2019)

IOP Conference Series: Materials Science and Engineering
Volume 795

Tomsk, Russia
28 October - 1 November 2019

ISBN: 978-1-7138-2671-2
ISSN: 1757-8981

Printed from e-media with permission by:

Curran Associates, Inc.
57 Morehouse Lane
Red Hook, NY 12571

Some format issues inherent in the e-media version may also appear in this print version.

Institute of Physics
Dirac House, Temple Back
Bristol BS1 6BE UK

Phone: 44 1 17 929 7481
Fax: 44 1 17 920 0979

techtracking@iop.org

Additional copies of this publication are available from:

Curran Associates, Inc.
57 Morehouse Lane
Red Hook, NY 12571 USA
Phone: 845-758-0400
Fax: 845-758-2633
Email: curran@proceedings.com
Web: www.proceedings.com

TABLE OF CONTENTS

MULTICOMPONENT ANTIFRICTION COMPOSITE BASED ON EXTRUDABLE MATRIX "UHMWPE - HDPE-G-VTMS - PP" FOR ADDITIVE MANUFACTURING 1
S V Panin, D G Buslovich, L A Kornienko, Yu V Dontsov

MICROSTRUCTURE AND MECHANICAL PROPERTIES OF NI-AL INTERMETALLIC THIN COATINGS PRODUCED BY MAGNETRON SPUTTERING 7
T S Ogneva, A A Ruktuev, D V Lazurenko, M Khomyakov, A Karmanova

SPH SIMULATION OF PLASTIC DEFORMATION IN HIGH VELOCITY IMPACT WELDING PROCESS OF 6061-T6 ALLOY PLATES 13
Yu Yu Emurlaeva, N S Aleksandrova, I A Bataev

MAIN PROBLEMS OF MATHEMATICAL MODELING HIGH ENERGIES PLASMA TECHNOLOGIES 20
V I Bogdanovich, M G Giorbelidze

MODELLING THE INTERACTION OF STATIONS IN THE PORT RAILWAY JUNCTIONS 28
E Borodina, E Prokofieva, S Vakulenko, A Petrov

THE WEAR RESISTANCE ADDITIVE PRODUCTS DERIVED FROM EED COBALT POWDER 39
E V Ageev, A Yu Altukhov, E V Ageeva

OBTAINING AND RESEARCH OF ADDITIVE PRODUCTS FROM ELECTROEROSIVE COBALT-CHROME POWDERS 45
E V Ageev, A Yu Altukhov, E V Ageeva

SIZE DEPENDENCE OF THE DISLOCATION SEGMENT VIBRATION SPECTRUM FOR VARIOUS TYPES OF DISLOCATIONS 50
I L Bataronov, V V Dezhin

RELATIONSHIP BETWEEN PHASE COMPOSITION AND WEAR RESISTANCE OF TITANIUM ALLOYS 55
I V Ivanov, A Thoemmes, K I Emurlaev, E I Tkachenko, E A Rudenya

NEURAL SIMULATION OF BALL MILL GRINDING PROCESS 61
A A Zakamaldin, A A Shilin

INFORMATION RESERVATION OF CYBER-PHYSICAL PRODUCTION OF THE INDUSTRY 4.0 67
D A Zakoldaev, A V Gurjanov, A V Shukalov, I O Zharinov

ESTIMATION OF STRESS-DEFORMED STATE OF THE CUTTING INSERT OF A DISK MILL USING COMPUTER-AIDED ENGINEERING 71
V G Shalamov, S D Smetanin, I S Boldyrev

ANALYSIS OF PRESSURE AND DYNAMIC TESTS OF THE BLOCKED WHEEL PAIR 79
V V Shiler, A V Shiler, I I Galiev, G I Petrov

STUDIES OF CURRENT MODULATION IN THE DC VOLTAGE LINK OF THE FREQUENCY CONVERTER DURING ITS OPERATION WITH ASYNCHRONOUS LOAD 85
O V Arsentiev, G O Arsentiev

MATHEMATICAL MODELING OF THE ELECTRIC DRIVE WITH UNIVERSAL POWER SEMICONDUCTOR CONVERTER ... 90
M P Dunaev, L V Arshinskiy

ORGANIZATION OF QUALITY CONTROL OF MANUFACTURED PRODUCTS IN SMART FACTORIES OF THE INDUSTRY 4.0 ... 98
D A Zakoldaev, A V Shukalov, I O Zharinov, O O Zharinov

METHOD OF CALCULATING OF SPIROID GEAR RESOURCE BY WEAR AT STEP LOADING MODE ... 103
A V Zaitsev

THE DEVELOPMENT OF DIGITAL TECHNOLOGIES IN MINING MACHINERY TECHNICAL MAINTENANCE ... 109
V I Klishin, B L Gerike, A A Mokrushev

ELECTROMAGNETIC METHOD FOR DETERMINING THE INTERNAL RESISTANCE OF CONDUCTIVE MATERIALS ... 117
F Y Kuznetsov

LOAD CAPACITY OF SPHERICAL ROLLER TRANSMISSION WITH DOUBLE-ROW PINION ... 122
M E Lustenkov, E S Lustenkova

IDENTIFICATION OF MATHEMATICAL MODEL OF DRYING UNIT ... 128
F P Perevoshchikov, N V Vuong, A K Shabo, A A Shilin

PLANETARY TORUS HELICAL TRANSMISSION ... 133
A P Prudnikov

INFLUENCE OF WEAR-RESISTANT COATING ON THE RELIABILITY OF REPLACEABLE INSERTS IN CUTTING STEEL ... 139
J Zhang, V N Kozlov, S Liu

DETERMINATION OF GEOMETRIC PARAMETER OF CYCLOIDAL TRANSMISSION FROM CONTACT STRENGTH CONDITION FOR DESIGN OF HEAVY LOADING MECHANISMS ... 145
E A Efremenkov, S K Efremenkova, I M Dyussebayev

POWER PARAMETERS AUTOMATED CALCULATION FOR TRANSMISSION WITH INTERMEDIATE ROLLING BODIES AND FREE CAGE ... 151
E A Efremenkov, E Bonnard

DEVELOPMENT OF ENERGY EFFICIENT MECHATRONIC MODULE FOR ALTERNATIVE ENERGY ... 157
K G Shibinskiy, E A Efremenkov, N E Galin

FACTORS INFLUENCING STRENGTH AND DIRECTION OF NATURAL VENTILATION PRESSURE IN GENERAL MINES TO CONTROL MINE AIRING ... 166
G B Lyalkina, A V Nikolaev, N S Makarychev

DESIGNING AN INFORMATION SECURITY SYSTEM IN SMART FACTORIES OF THE INDUSTRY 4.0 ... 171
D A Zakoldaev, A V Shukalov, I O Zharinov, O O Zharinov

Author Index

IOP Conf. Series: Materials Science and Engineering **795** (2020) 011001 doi:10.1088/1757-899X/795/1/011001

Preface

The International Conference on Mechanical Engineering and Modern Technologies 2019 (MEMT2019) was held at Tomsk Polytechnic University, Tomsk. Russia on October 28–November 1, 2019. MEMT2019 is a scientific forum where researchers, students and industry professionals exchange their ideas, share experience and discuss technical issues in the field of mechanical engineering, material sciences, modelling and automation. The conference program included as well the workshop for young scientists "Modern technologies and equipment" with participation of representatives of Ariel Corporation, USA. The representatives of such companies as Komatsu (Germany) and Domist (Czech Republic) spoke at the Conference opening ceremony. The sessions covered the following topics:

- Machine and mechanisms reliability;
- Electrical engineering and mechatronics;
- New manufacturing technology; Materials science and nanotechnology in mechanic engineering;
- Automation and modeling of engineering processes;
- Industrial safety;
- Design in mechanic engineering.

We appreciate contributions from all participants of MEMT2019. We also want to thank sponsors for the financial support and Program Committee members for their efforts in organization and holding of the MEMT2019 Conference.

Dr. Evgeniy N. Pashkov,

Chair of MEMT2019 Organizing Committee

Tomsk Polytechnic University

IOP Conf. Series: Materials Science and Engineering **795** (2020) 011002 doi:10.1088/1757-899X/795/1/011002

Peer review statement

All papers published in this volume of *IOP Conference Series: Materials Science and Engineering* have been peer reviewed through processes administered by the proceedings Editors. Reviews were conducted by expert referees to the professional and scientific standards expected of a proceedings journal published by IOP Publishing.

Multicomponent Antifriction Composite Based on Extrudable Matrix "UHMWPE - HDPE-g-VTMS - PP" for Additive Manufacturing

S V Panin[1,2,a], D G Buslovich[1,2], L A Kornienko[1] and Yu V Dontsov[2]

[1] Institute of Strength Physics and Materials Sciences SB RAS, 634055, Tomsk, Russia
[2] National Research Tomsk Polytechnic University, 634050, Tomsk, Russia

E-mail: [a] svp@ispms.tsc.ru

Abstract. Tribomechanical properties of antifriction composites based on the extrudable matrix "UHMWPE + 17 wt. % HDPE-g-VTMS + 12 wt. % PP" with chopped fiberglass formed by three methods: Hot Pressing of Powders (HPP), Hot Pressing of Granules (HPG) and Fused Deposition Modeling (FDM) was studied. It has been found that a composite fabricated by the FDM method possesses the highest strength properties (elastic modulus, yield strength and tensile strength). It is shown that tribological properties (friction coefficient, volumetric wear) of composites fabricated by the three methods are close to each other that is related to impact of the reinforcing filler (fiberglass). The latter takes on compressive and shearing loads during tribo-loading and improves wear resistance of the composite. The studied multicomponent UHMWPE based composite is recommended for use as a feedstock for the manufacturing antifriction products by additive manufacturing.

1. Introduction

Currently, applied studies on the use of 3D printing methods, namely FDM (Fused Deposition Modeling) for manufacturing products with polyolefin matrices (including UHMWPE) using various fillers and plasticizers [1-6] are carried out.

Besides the market available additives traditionally used in industry, some commercially available polymers compatible with UHMWPE can be used as plasticizing fillers. Among them grafted high-density polyethylene HDPE-g-VTMS and HDPE-g-SMA might be mentioned. Loading of such polymeric plasticizers ensures to a blend an acceptable level of melt flow index. This makes it possible to produce extrudable polymer-polymeric composites with UHMWPE matrix with specified strength and wear resistance characteristics for manufacturing products by 3D printing technologies.

According to [7] satisfactory melt flow index of UHMWPE blends can be achieved when the content of plasticizing additives makes at least 20 wt. %. This ensures maintenance of the spherulitic pattern of permolecular structure of the matrix (when material is fabricated by compression sintering).

On the basis of experimentally obtained results an algorithm for determining control parameters (composition) at minimum experimental data available is proposed. The latter gives the required (restrictive) values to the effective properties of polymer-polymeric composites. On the basis of the specified i) composition structure, ii) tribological and iii) mechanical properties of polymer-polymeric UHMWPE composites experimental specimens were fabricated by the FDM (extrusion layer printing) and studied.

It is shown that tribomechanical properties of extrudable polymer-polymeric UHMWPE based composites fabricated by 3D printing surpass the similar properties compared to composites fabricated by compression sintering. The reason of the latter might be associated with the formation of more homogeneous permolecular structure primarily due to the employing a twin-screw compounding technique.

In the present work, some extrudable multicomponent polymer-polymeric UHMWPE composites filled with glass-fibers (200 μm long) have been fabricated. Their tribotechnical properties are studied with a view application in friction units of machines and mechanisms operating under various tribo-loading conditions.

2. Experimental

Powder of ultra-high molecular weight polyethylene (UHMWPE) by Ticona (GUR-2122) with a molecular weight 5.0 10^6 and particle size of 5–10 μm was used as the matrix resin. Grafted high-density polyethylene HDPE-g-VTMS (ground granulate), as well as polypropylene powder grade P21030 (MFR = 3.0 g/10 min) were used as plasticizing agents. Milled glass fibers 200 μm long were used as a filler (supplier "Graphite Pro" LLC, Moscow) at the content of 5 wt. %. Mixing of the matrix powder and fillers for specimen fabrication by Compression Sintering (Hot Pressing) was carried out in a planetary ball mill MP/0.5·4 (Tekhnocentr ltd., Rybinsk) with preliminary dispersion of the components in an ethanol suspension in an ultrasonic bath PSB-Hals 1335-05-(Gals ltd., Moscow).

In order to efficiently disperse fine UHMWPE particles (units of microns) with coarse particles of the polymer fillers (hundreds of microns) they were compounded in a twin-screw extruder Rondol (10 mm Twin Screw Extruders, Microlab). The temperature at the outlet extrusion nozzle was T=210 °C. Granules (with an average size of 3 – 5 mm) were obtained by subsequent mechanical grinding of the extrudate with the use of a Rondol shredder.

Rectangular shape samples of the polymer composites with the size 65×70×10 mm were fabricated by i) hot pressing of three-component powder mixtures at the specific pressure 10 MPa and the temperature of 200 °C with the use of a laboratory unit based on a hydraulic press "MS-500" (LLC SPC "TechMash" ltd.) equipped with an split ring furnace; the subsequent cooling rate made 5 °C/min; ii) FDM (Fused Deposition Modeling) from granules of the same polymer blend with the use of a laboratory 3D–printer ArmPrint – 2 (National Research Tomsk Polytechnic University) with a nozzle diameter of 0.4 mm; the temperature of the stage and upper and lower filament (granules) feeding regions was T = 90, 160 and 200 °C, respectively. The layer-by-layer deposition rate and the thickness of the applied layer of the material were 20 mm / s and 0.3 mm, correspondently. Specimens of the required shape and size were cut out from the fabricated tiles with the help of a computer-controlled vertical milling machine.

Mechanical properties were determined under tensile tests with the use of electromechanical testing machine "Instron 5582". Dog-bone shape specimens were prepared for testing with the number of each type not less than 4.

The volumetric wear of specimens under the dry sliding friction was determined by the "pin-on-disk" scheme with the use of CSEM CH2000 tribometer (CSEM, Switzerland) at the load of P = 5 N (calculated contact pressure P_{max} = 31.8 MPa) and sliding velocity V = 0.3 m/s. The diameter of the bearing GCr15 steel ball counterface (HRC 60) was 6 mm.

The observation of the wear tracks surface topography was carried with the use of Neophot 2 optical microscope (Carl Zeiss Jena, Germany) equipped with a digital photo camera Canon EOS 550D (Canon Inc., Japan) as well as a stylus profilometer Alpha-Step IQ (KLA-Tencor). Structural studies were performed with the help of scanning electron microscope LEO EVO 50 (Carl Zeiss, Germany) at the accelerating voltage of 20 kV on the rupture surface of notched specimens failed after preliminary exposure in liquid nitrogen. The crystallinity was measured with the use of a combined analyzer SDT Q600 (TA Instruments, USA).

3. Results and discussion

Mechanical properties of neat UHMWPE, multicomponent mixture "UHMWPE + 17 wt. % HDPE-g-VTMS +12 wt. % PP″ and their composite fabricated by all three methods are shown in table 1. It is seen that mechanical properties (elastic modulus, yield strength, tensile strength) of the composite fabricated by 3D-printing exceeds those of the composite prepared by powder hot pressing (HPP) by 10 %.

SEM micrographs of permolecular structure of the thermoplastic three-component mixtures ″UHMWPE + 17 wt. % HDPE-g-VTMS +12 wt. % PP″ and composites based on this matrix fabricated by granulate sintering (HPG) and the FDM are presented in Figure 1. It is seen from the figure that loading of the fiberglass at granules hot pressing (HPG) ensures more uniform permolecular structure. The latter is formed first of all due to twin-screw compounding (Figure 1, c and d). This provides an increase of key mechanical properties (Table 1).

Table 1. Mechanical properties of UHMWPE and its composites with loading 30 wt. % of polymer plasticizers (total) as well as additionally filled with GF.

Filler content, wt. %	Density, g/cm^3	Shore D hardness	Elastic modulus E, MPa	Yield point $\sigma_{0.2}$, MPa	Ultimate strength σ_U, MPa	Elong. at break ε, %
UHMWPE	0.934	57.7±0.6	711±40	21.6±0.6	42.9±3.1	485±28
″UHMWPE +17 wt. % HDPE-g-VTMS +12 wt.% PP″ (powder blend, HPP)	0.933	57.8±0.3	876±71	25.1±0.3	19.1±1.3	149±38
- Granulate, HPG	0.939	59.1±0.2	907±75	26.8±0.6	34.3±2.7	414±69
- Granulate, FDM	0.926	57.9±0.4	922±58	25.6±0.4	30.3±0.7	356±63
″UHMWPE +17 wt. % HDPE-g-VTMS +12 wt.% PP + 5 wt.% GF″ (powder blend, HPP)	0.947	59.9±0.6	1303±99	24.1±1.9	20.9±0.5	49±11
- Granulate, HPG	0.945	60.4±0.6	1093±83	26.5±0.9	25.3±3.2	228±68
- Granulate, FDM	0.929	61.5±1.2	1376±71	29.3±0.3	29.9±2.1	306±19

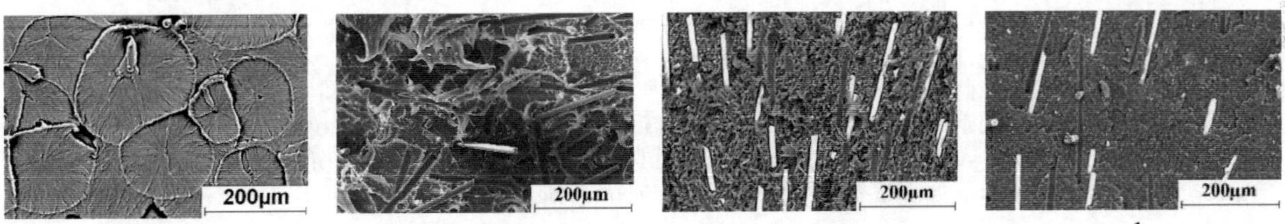

a b c d

Figure 1. SEM- micrographs of permolecular structure of UHMWPE (a) and its composites ″UHMWPE+17 wt. % HDPE-g-VTMS + 12 wt. % PP + 5 wt. % GF″ (b- powder blend, HPP; c - Granulate, HPG; d- Granulate, FDM).

Further, the tribological properties of the composites, including ones filled with the fiberglass were studied under dry sliding friction. The results are shown in table. 2.

Table 2. Data on friction coefficient and volumetric wear of UHMWPE based composites loaded with 5 wt. % of fiberglass ("pin-on-disk" scheme).

№	Filler content, wt. %	Wear rate, mm^3/h	Friction coeff., f

1	UHMWPE	0.102±0.003	0.136±0.024
2	″UHMWPE +17 wt. % HDPE-g-VTMS +12 wt.% PP″ (powder blend, HPP)	0.093±0.003	0.117±0.007
3	- Granulate, HPG	0.095±0.003	0.132±0.009
4	- Granulate, FDM	0.096±0.004	0.112±0.029
5	″UHMWPE+17 wt. % HDPE-g-VTMS +12 wt.% PP + 5 wt. % GF″ (powder blend, HPP)	0.073±0.005	0.098±0.003
6	- Granulate, HPG	0.075±0.004	0.092±0.004
7	- Granulate, FDM	0.079±0.005	0.103±0.013

It is seen in the table 2 that loading of fiber glasses up to 5 wt. % into thermoplastic matrix «UHMWPE + 17 wt. % HDPE-g-VTMS +12 wt. % PP» gives rise to decreasing volumetric wear and friction coefficient of composites formed by all three methods (HPP, HPG and FDM). Moreover, their values are close to each other regardless of the method of composites' fabrication. This is clearly seen in Figure 2, a, b as well.

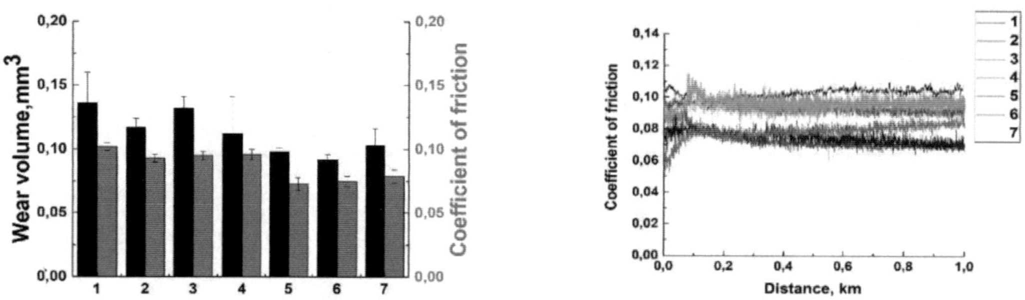

a b

Figure 2. Volumetric wear (a) and coefficient of friction (b) for UHMWPE (1) and its composites: ″UHMWPE + 17 wt. % HDPE-g-VTMS + 12 wt. % PP″ formed by the HPP (2), HPG (3), FDM (4) ″UHMWPE + 17 wt. % HDPE-g- VTMS + 12 wt. % PP + 5 wt. % GF″ formed by the HPP (5), HPG (6), FDM (7).

In other words, micron-sized glass fibers in a multicomponent composite with the extrudable UHMWPE matrix effectively take on compressive and shear loads during triboloading and protect the matrix from wear. This is supported by the observed topography micrographs of wear track surfaces of fiberglass-reinforced composites fabricated by all three methods (HPP, HPG and FDM) (Figure 3).

1 2 3

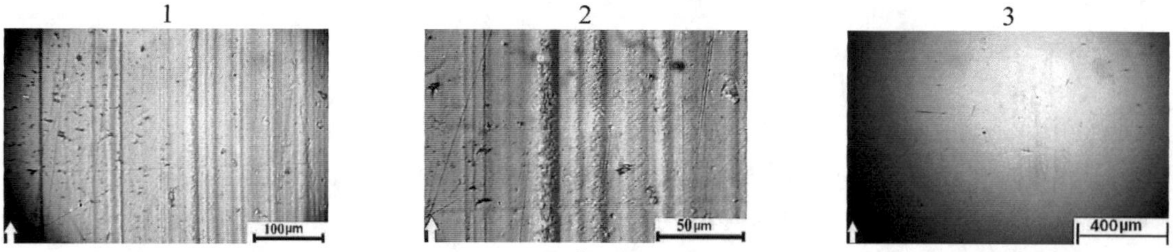

a b c

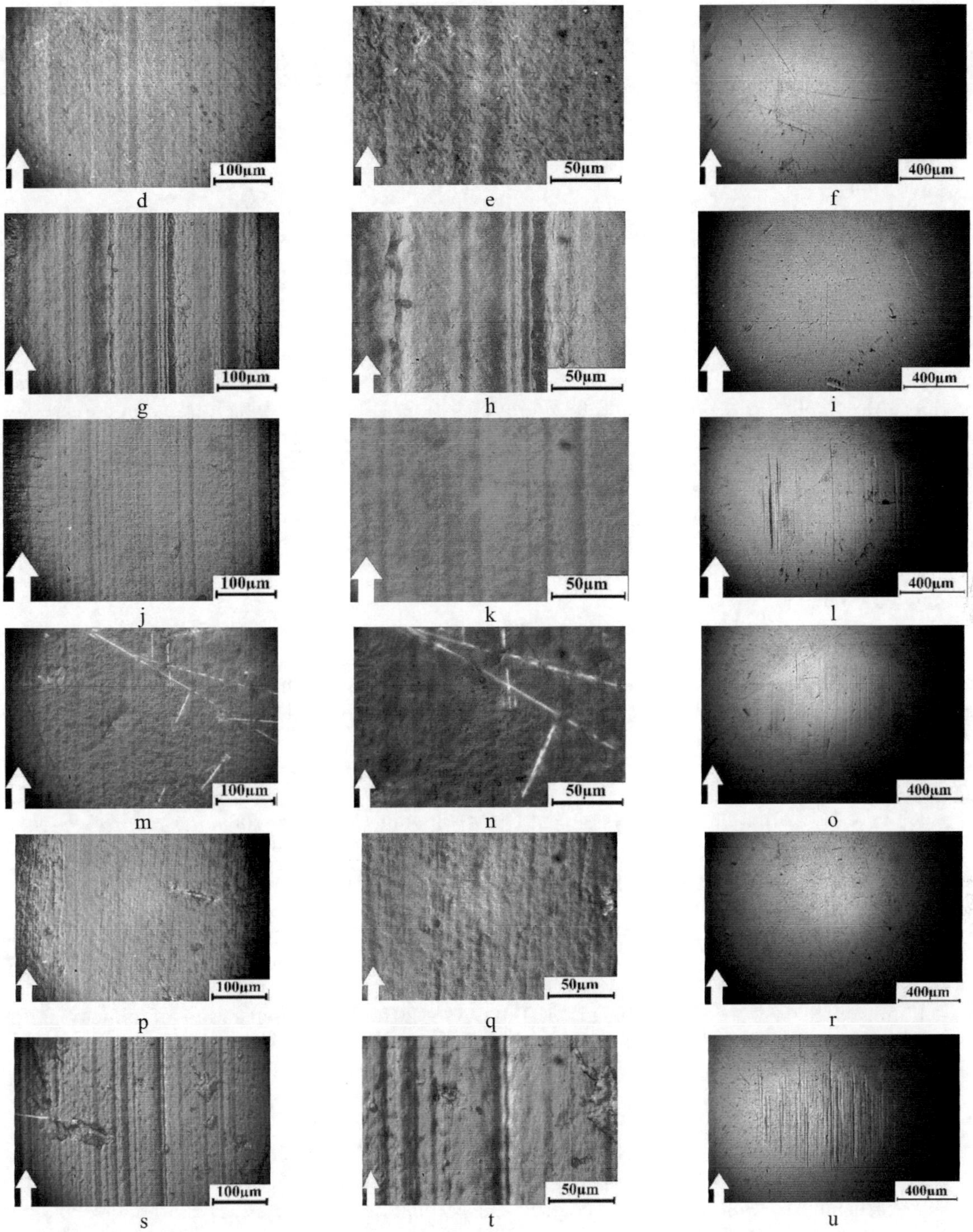

Figure 3. Optical images of wear track surface topography of polymer composites (1, 2) and bearing steel counterface (3): UHMWPE (a, b, c) and polymer composites "UHMWPE + 17 wt. % HDPE-g-VTMS + 12 wt. % PP″ fabricated by HPP (d, e, f), HPG (g, h, i), FDM (j, k, l); ″UHMWPE + 17 wt. % HDPE-g-SMA + 12 wt. % PP + 5 wt. % GF″ obtained by HPP (m, n, o), HPG (h, g, r), FDM (s, t, u). Dry sliding friction; V=0.3 m/s and P= 5 N ("pin-on-disk"; CSEM CH2000 tribometer).

Data of figures 2 and 3 that under moderate triboloading conditions (V = 0.3 m/s and P = 5 N) a multicomponent extrudable UHMWPE based composite with the GF content of 5 wt. % possess quite high wear resistance. It means that this composite can be processed both at standard manufacturing methods (screw extrusion, injection moulding, etc.), as well as for FDM-printing for fabricating complex shape products for friction units in mechanical engineering.

4. Conclusion

Tribomechanical properties of antifriction composites based on the extrudable matrix ″UHMWPE + 17 wt. % HDPE-g-VTMS + 12 wt. % PP″ loaded with milled fiberglass formed in three ways: hot pressing of powders (HPP), hot pressing of granules (HPG) and layer-by-layer extrusion printing of granules (FDM) have been investigated.

It has been shown that the highest strength properties (elastic modulus, yield strength and tensile strength) are possessed by a composite fabricated by the FDM method.

The tribological properties (friction coefficient, volumetric wear) of composites fabricated in three ways are close to each other that is related to the reinforcing filler (fiberglass) which takes on compressive and shear loads during tribo-loading and protects the matrix from wear. The reason might be also related to use tribotesting scheme "pin-on-disk" when a highly localized loading of ductile elastic polymer is not accompanied with high wear rate.

The studied multicomponent composite based on UHMWPE is recommended as a feedstock to manufacturing anti-friction products by FDM printing.

Acknowledgment

This work was performed within the frame of the Fundamental Research Program of the State Academies of Sciences for 2013-2020, line of research III.23. This study was financially supported by the grant of the President of the Russian Federation for state support of the leading scientific schools of the Russian Federation SSh-5875.2018.8. The reported study was funded by RFBR, project number 19-38-90106.

References

[1] Kurtz S 2016 High Pressure Crystallized UHMWPEs in *UHMWPE Biomaterials Handbook: 3-rd Edition* ed. S Kurtz (Norwich, NY,William Andrew Publishing) pp. 434-448.

[2] Borges R A, Choudhury D, Zou M 2018 3D printed PCU/UHMWPE polymeric blend for artificial knee meniscus *Tribol. Int.* **122** pp 1–7

[3] Bala A S, Wahab M S, Ahmad M, Soon C F, and Ramli M S 2016 Processability and Thermal Properties of Ultrahigh Molecular Weight Polyethylene/Polypropylene Blends *ARPN Jour. of Engin. and Appl. Sci.* **11** (8) pp 5481–86.

[4] Ramli M S, Wahab M S, Ahmad M, Bala A S 2016 FDM preparation of bio-compatible UHMWPE polymer for artificial implant *Int. J. Adv. Manuf. Technol.* **11** (8) pp 5473-5480

[5] Changhui S, Aibing H, Yongqiang Y et al. 2016 Customized UHMWPE tibial insert directly fabricated by selective laser sintering *Int. J. Adv. Manuf. Technol.* **85** pp 1217–1226

[6] Gongde Liu, Huilin Li 2003 Extrusion of ultrahigh molecular weight polyethylene under ultrasonic vibration field *J. Appl. Polym. Sci.* **89** (10) pp 2628–2632

[7] Panin S V, Buslovich D G, Kornienko L A, Alexenko V O, Dontsov Yu V, S.V. Shil'ko 2019 Structure, as well as the Tribological and Mechanical Properties, of Extrudable Polymer-Polymeric UHMWPE Composites for 3D Printing *J. Frict. Wear* **40** pp 107–115

IOP Conf. Series: Materials Science and Engineering **795** (2020) 012002 doi:10.1088/1757-899X/795/1/012002

Microstructure and mechanical properties of Ni-Al intermetallic thin coatings produced by magnetron sputtering

T S Ogneva[1], A A Ruktuev[1], D V Lazurenko[1], M N Khomyakov[2] and A E Karmanova[1]

[1] Materials Science Department, Novosibirsk State Technical University, 20 Karl Marks ave., Novosibirsk 630073, Russian Federation

[2] Laser Plasma Department, Institute of laser physics SB RAS, Acad. Lavrentiev ave., Novosibirsk 630090, Russian Federation

E-mail: pandorra.06@mail.ru

Abstract. In the present study the thin NiAl intermetallic foils formed on different types of substrates by magnetron sputtering technique were investigated. To provide the deposition of intermetallic NiAl compound in one step without additional heat treatment the composite targets assembled from parts of Al and Al plates were used. The structure of formed thin NiAl coatings was studied using scanning electron microscopy and X-ray diffraction analysis. Mechanical properties were assessed by nanohardness measurements and wear testing of deposited coatings. During sputtering the distance from the target to the substrates varied from 60 to 100 mm to estimate the effect of this parameter on structure and properties of the coatings. The results revealed that thin coatings sputtered at the closer distance from the target to the substrates had the higher hardness about 11 GPa and exhibited the high level of wear properties.

1. Introduction

In the present time, intermetallics are known as effective materials for protective coatings of metals and alloys used in aircraft and rocket building, microelectronics, chemical and petroleum industries. Among all intermetallic compounds nickel aluminides are of particular interest, especially Ni_3Al and NiAl, whose main advantages are a relatively high melting point (1385–1638 °C), a relatively low density (5–7.5 g/cm^3), a high elastic modulus (190–290 GPa), resistance to creep. The NiAl phase is characterized by high wear resistance and resistance to oxidation (including elevated temperatures), which makes it attractive for use as protective heat-resistant and wear-resistant coatings. Today, nickel aluminides are actively used as coatings for gas turbine engines based on nickel superalloys and stainless steel products, as sublayers for magnetic recording devices, surface catalysts, and thermistor thin films in microelectronics. Nevertheless, the problem of the brittleness of intermetallic materials still remains the topical. Besides, it is necessary to solve the problems of improving the coating to the substrate, increasing the resistance to oxidation and tribotechnical properties.

Magnetron sputtering is the technique which allows depositing coatings formation with high density of micro- or nanocrystalline structures, provides the complete absence of the capillary phase,

[1] To whom any correspondence should be addressed.

possibility of cladding the coatings on the thermosensitive materials at low temperatures, high deposition rate, and high mechanical and functional properties of formed coatings.

There are two approaches to form intermetallic coatings by magnetron sputtering technique. The first is the simultaneous deposition of two or more metallic targets in a chamber with two magnetrons. The second is the preliminary manufacturing of a multicomponent target corresponding to the desired composition. In the case with simultaneous deposition from two or more targets, it is problematic to vary the number of knocked out atoms, so it is difficult to deposit the material of a given composition. In addition, devices with several magnetrons are quite expensive. When manufacturing multicomponent targets, as a rule, hot isostatic pressing or spark plasma sintering are used. These technologies are quite and time-consuming processes, which causes additional material costs. The feature of the proposed approach is the use of composite targets, which is assembled from alternating blocks of two materials. The number and width of Ni and Al parts in the composite target is chosen based on the composition of the coating to be obtained. Such approach will allow quickly and efficiently deposit thin layers with a given composition, and allows rationally use the capabilities of simple sputtering device equipped with the only one magnetron.

Thin coatings and thin foils were obtained and studied on the basis of various combinations of materials: Ni-Al, Ni-Al-X (where X = N, Cu, W, Ti, etc.) [1-5], Ti-Al , Ti-Al-X (X = C, Cu, Ni, Co, Fe, Cr, Nb) [6-8], Fe-Al, Ni-Ti, AlNi-Al_2O_3, Al-Au [9], Al-Zr [10]. The main methods for studying thin intermetallic coatings are scanning and transmission electron microscopy and X-ray phase analysis. Due to the small thickness of the coatings, the strength properties are mainly evaluated by the method of nanoindentation.Magnetron sputtering of nickel and aluminum with subsequent heat treatment showed that fining of the structure to nanometer sizes reduces the temperature of intermetallic formation to below the melting point of aluminum (190–250 °C) [1, 2, 11]. At the moment, a lot of scientists are trying to increase the adhesion of intermetallic coatings based on nickel and aluminum to the substrate, to eliminate the delamination during thermal cycling, to avoid the diffusion of coating materials in the substrate during heating, to eliminate increase of tribotechnical properties. Such problems can be partly solved by doping other elements, by fining the nanostructure, by varying the modes of magnetron sputtering and using different heat treatment modes. [2, 7, 12–15].

The author found the only publication [15], in which target consisting of concentric parts (inside part - a circle of nickel, outside part - a ring of aluminum) were used. Nevertheless, structural features of nanostructure and phase analysis weren't presented enough in details. Thus, the purpose of the present study is to evaluate the structure and mechanical properties of intermetallic coatings sputtered from composite targets assembled from Ni and Al parts on metallic and non-metallic materials in various sputtering conditions.

2. Materials and methods

The target consisted of alternating bars (strips) of cp-nickel and cp-aluminum cut from plates on wire electrical discharge machine Sodik AG400L. Diameter of the target was 75 mm, the width of side Ni and Al plates was 17.5 mm, the width of central was 10 mm, the thickness of the target was 4 mm. Cp-titanium, steel (0.2 wt. % C), Al_2O_3 (doped with spinel $MgAl_2O_4$) and silicon monocrystallic plates were used as substrates. Magnetron sputtering experiments were carried out on magnetron sputtering device PVD-DESK-PRO. The base pressure of vacuum system was reduced to 5·E-5 prior to sputtering. During the deposition process in the atmosphere of high purity Ar (99.999%) a pressure was maintained at 5 – 5.4 ·E-3 hPa. The applied DC power was 600 W. In the study the varied parameters were the distance between table and magnetron, and magnetron power (table 1).

The micro- and nanostructure of deposited coatings on the top of the surface and in the cross section were investigated on scanning electron microscope (SEM) Carl Zeiss EVO 50 XVP, with the scale ×1000… ×300000 and accelerating voltage 3–10 keV. X-ray analysis were carried out in θ-θ diffractometer ARL X'TRA using Cu $K_{\alpha1+\alpha2}$ radiation with the scanning step 2θ = 0.05° and dwell time 3…10 pps. Nanohardness measurements and scratch tests were conducted on scanning nanohardness tester Naniscan-3D equipped with atomic force microscope in Institute of Laser Physics

(Novosibirsk). The hardness of the material was determined by penetration depth of Berkovich diamond pyramid with the loading increased from 0.1 to 50 mN). The loading during scratch tests were varying in the same range. The coatings were scratched in two different ways: face forward and edge forward orientation of the indentor. Scratch tests provide information about maximum loading at which plastic deformation of the material and delamination of the coating begins. The tribological properties were estimated on by wear test Naniscan-3D using Berkovich indentor at the loading 20 mN, the number of cycles was 500.

Table 1. Parameters of magnetron sputtering experiment.

Material of the substrate	Distance from the target to the substrate (h), mm	Duration of sputtering, min	Power, W
Al2O3, Si	100	30	600
Al2O3, Si	85	30	600
Al2O3, Si	60	30	600
Al2O3, Si	100	2	600
Ti	60	30	600
Steel	60	30	600

3. Results and discussion

3.1. Structure of deposited coatings

SEM investigations of fracture surface in the cross section of the samples showed that the coatings has columnar nanostructure, which is typical for materials obtained by magnetron sputtering technique (figure 1). The intermetallic layer near the substrate is the smallest and has equiaxial shape. Further increase in the film thickness leads to the formation of a columnar structure that is explained by the Volmer-Weber film growth mechanism, where at the initial stage equiaxed nanoscale crystallization centers nucleate, and then they increase and form the columnar structure. The appearance of fracture surface indicates the brittle fracture of intermetallic layers. The coating thickness increases with decrease of the distance from the target to the substrate and reaches 1.8 μm, 3.2 μm and 5μm at 100, 85 and 60 mm, respectively. The morphology of the deposited surface at different height of sputtering significantly changes. On figure 2 the top surface of coatings deposited on Si substrates are presented. At the distance 100 mm the velocity of sputtered depositing atoms is low, thus, the crystals grow slowly and nothing disturb their formation and growing of clusters at the top layer. When the substrate is situated closer to the target and to the source of Ar atoms (h = 60 mm) the top surface is densifying by just accelerated atoms from the target, besides, concentration of Ar atoms is greater at this distance, therefore, some atoms can deform the surface when randomly interacting with it.

3.2. X-Ray analysis

The samples deposited during 30 minutes were used for X-ray analysis because of thickness of the obtained surface. Phase identification revealed the presence of the only intermetallic phase NiAl, peaks of initial metallic components presenting in the target weren't found (figure 3). Thus, the suggestions can be made, that the reaction with formation of intermetallic phases goes after the deposition on the atoms due to the heating of the substrate.

Reflections of NiAl (111) planes show great intensity on X-Ray diffraction patterns that indicates strong crystallographic texture that isn't typical for this phase. Formation of textured crystal structure in the coatings can be possibly connected to the columnar atom-to-atom mechanism of its growing. As it was described before, grains forming on the substrate at the first stage of the film growing are mostly equiaxial At the distance 100 mm growing rate of the coatings isn't very high, thus, equiaxial grains takes most part of the coating. That is why the relative intensity correlates with powder diffraction pattern of the NiAl phase. With increase of growing rate of the coatings at closer distance (h = 85 and 60 mm) columnar oriented crystal growing becomes predominant, and at the distance h=60 mm in deposited coatings almost all crystallographic planes except (111) degenerate, which is represented in the diffraction pattern — there is only one, but extremely intense, reflection of (111) phase (figure 3 a).

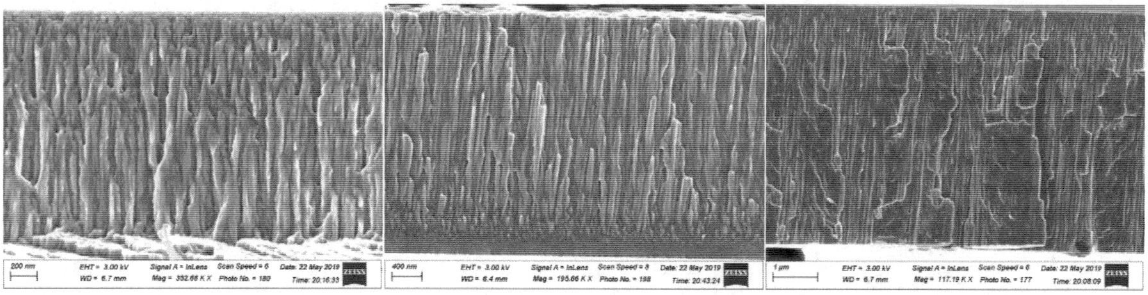

Figure 1. Microstructure of thin intermetallic NiAl coatings in the cross section sputtered at the distance from the target to substrate (from left to right): 100 mm, 85 mm, 60 mm.

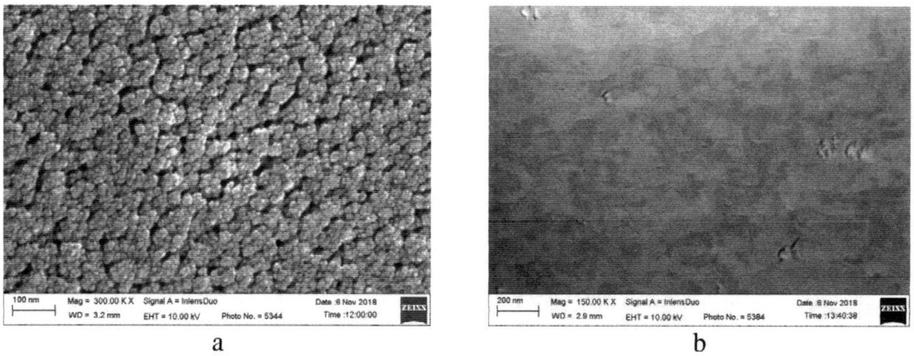

Figure 2. Microstructure of the top surface of thin intermetallic NiAl coatings sputtered at the distance 100 mm (a) and 60 mm (b).

Crystallographic texture in this case depends on growing rate of planes with different crystallographic orientation. Besides, the structure of substrates can also influence on the orientation of as deposited crystals. The phase composition of coatings sputtered on different substrates (steel, titanium, alumina and silicon) is the same, the X-ray diffraction revealed only NiAl phase in deposited coatings (figure 3 b). Nevertheless, X-ray reflexes of intermetallic coatings on metallic substrates show more uniform distribution of its intensity. The intensity of (111) planes in the coatings sputtered on metallic substrates is relatively low, besides, the reflexes of planes (110) and (211) appears which wasn't found in X-ray diffraction patterns of coatings sputtered on silicon and alumina plates. This can be explained by different roughness and surface morphology of the substrates, the quality of preparation of substrates and the physical properties of the substrate materials.

3.3. Nanohardness measurements

The nanohardness of intermetallic coatings reaches the highest values ~10 GPa in the samples sputtered at the distance 100 mm. Measurement results showed that nanohardness value slightly grows with when the distance from the target to the substrates becomes less. According to SEM results, such behavior can be connected to aforementioned differences in structure of surfaces.

IOP Conf. Series: Materials Science and Engineering **795** (2020) 012002 doi:10.1088/1757-899X/795/1/012002

It also should be noted that nanohardness of NiAl coatings deposited at the distance 60 mm on metallic substrates reaches 12 GPa and higher, when coatings sputtered on non-metallic (Al2O3 and silicon) substrates exhibited the level of nanohardness about 10 GPa. Basing on X-ray diffraction investigations, such behavior of nanohardness level can be connected to anisotropy of mechanical properties in thin layers with difference crystallographic orientation.

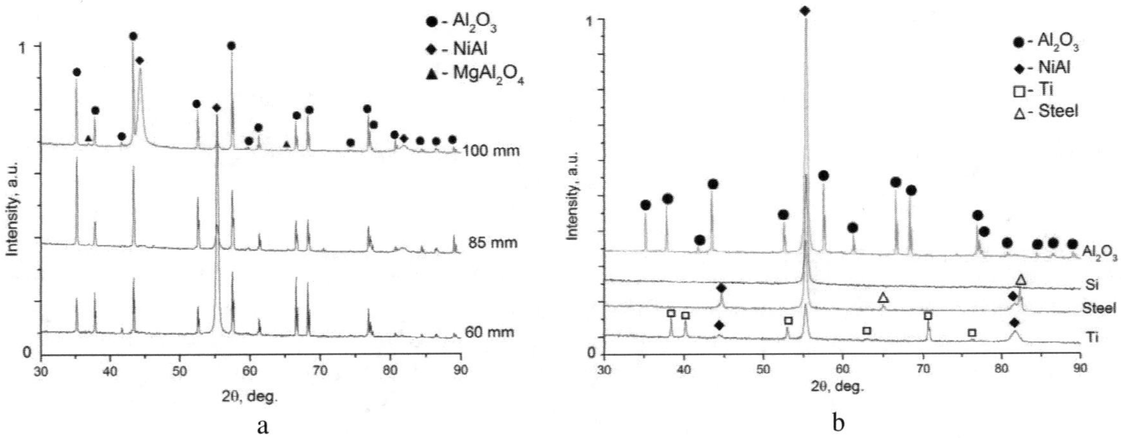

Figure 3. X-ray diffraction patterns of the coatings sputtered on ceramic substrates at different distance from the target (a); sputtered on different substrates at the distance 60 mm (b).

3.4. Scratch and wear tests

Samples with the thin layers sputtered during 2 min at the distance 100 mm and power 700 W were used to estimate the adhesion behavior of intermetallic coatings by scratch tests. Adhesion behavior war estimated on samples sputtered during 2 min, because in this case the maximum loading is enough to penetrate the entire thickness of the coatings. To carry out wear tests, the samples sputtered during 30 min were chosen, because their grater thickness will allow much number of cycles.

AFM image of the scratched sample presented on figure 4 a. Profile 1 on figure 4 b corresponds to edge-forward movement of the pyramidal indenter (reproduces micro-cutting of material); profile 2 – face-forward movement (represents plastic extrusion).

As it can be seen from the measurement results, the indenter penetration depth into the samples during scratching reaches 0.25–0.30 μm. The coating thickness on these samples, measured on interferometer, reaches 0.22 μm. Thus, the conclusion can be made that in the process of testing the indenter penetrates the entire depth of the coating, and also enters into the substrate material. Images of sample profiles show that there are no peeling and chips along the entire length of the scratches both on metallic and silicon substrates that can be the evidence of good adhesion of intermetallic coatings to the substrate material.

The results of wear tests showed that substrates didn't broke or delaminated after 500 cycles of wearing by Berkovich indentor at the loading 20 mN. The coatings sputtered at the distance 60 mm exhibited the best wear results, depth of wear crater after 500 cycles was ~500 nm. The samples deposited at 85 and 100 nm showed more intensive wearing; the depth of the tracks after 500 cycles was ~600 nm. The obtained results correlate with microhardness measurements and also can be an effect of slight densifying of the intermetallic layer at the close distance to the target. The nanohardness and wear properties of obtained coatings are comparable and even exceed the level of NiAl coatings obtained at similar conditions [15].

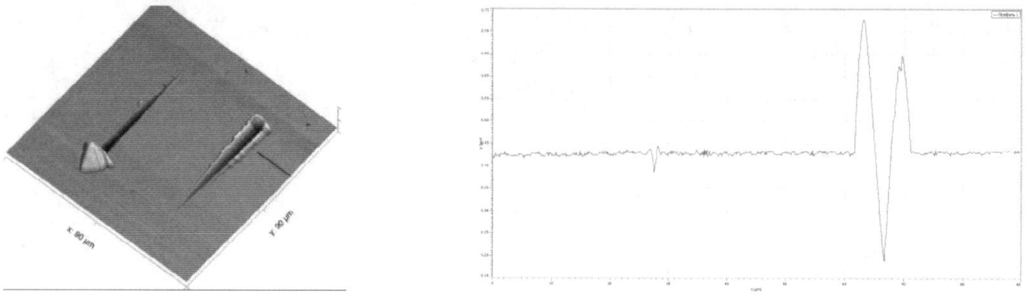

a b

Figure 4. AFM image of the tracks after the scratching tests (a), profile measured by the black line on AFM image (b).

4. Conclusions

As a result of the carried out investigations the following conclusions can be drown. Magnetron sputtering of Ni and Al of the composite target assembled from the parts of elemental blocks of pure initial metals allows deposition of intermetallic thin coatings containing only NiAl intermetallic phase. Magnetron sputtering promotes the growth of columnar grains with (111) plane preferable orientation. The morphology of the surface and grain size depends on the distance between the target and the substrate; in particular, moving the substrate closer to the target leads the densification of the depositing surface and increase of the depositing rate. This fact influences on nanohardness level, which grows from 10 to 11 GPa with the reduction of the distance from target to the substrates, and affects the wear characteristics, which increase by 20 %.

Acknowledgments

This work was carried out with the financial support by Ministry of Science and Education of Russian Federation (Grant of the President No. MK-5849.2018.3, Agreement No. 075-15-2019-665).

References

[1] Simoes S, Viana F, Ramos A S, Vieira M T and Vieira M F 2001 *Intermetallics* **19** 350
[2] Noro J, Ramos A S, Vieira M T 2008 *Intermetallics* **16** 1061
[3] Espinoza Torres C, Condó A M, Haberkorn N, Zelaya E, Schryvers D, Guimpel J and Lovey F C 2014 *Mater. Charact.* **96** 256
[4] Liu K T and Duh JG 2008 *Surf. Coat. Technol.* **202** 2737
[5] Li P Y, Lu H M and Tang S C and Meng X K 2009 *J. Alloys Compd.* **478** 240
[6] Shaginyan L R, Gorban V F, Krapivka N A, Firstov S A, and Kopylov I F 2016 *J. Superhard Mater.* **38** 25
[7] Zhang Ch, Feng K, Li Zh, Lu F, Huang J, Wu Y and Chu P K *Mater. Des.* **111** 575
[8] Zhong D, Moore J J, Sutter E and Mishra B 2005 *Surf. Coat. Technol.* **200** 1236
[9] Moser M, Mayrhofer P H, Ross I M and Rainforth W M 2007 *J. Vac. Sci. Technol., A* **25(5)** 1402
[10] Gudla V C, Bordo K, Engberg S, Rechendorff K and Ambat R 2016 *Mater. Des.* **95** 340
[11] Edelstein A S, Everett R K, Richardson G Y, Qadri S B, Altman E I, Foley J C and Perepezko J H 1994 *J. Appl. Phys.* **76** 7850
[12] Duarte L I, Ramos A S, Vieira M F, Viana F, Vieira M T and Koçak M 2006 *Intermetallics* **14(10-11)** 1151
[13] Devia D M, Restrepo-Parra E, Arango P J, Tschiptschin A P and Velez J M 2011 *Appl. Surf. Sci.* **257(14)** 6181
[14] Tang C, Klimenkov M, Jaentsch U, Leiste H, Rinke M, Ulrich S, Steinbrück M, Seifert H J and Stueber M 2017 *Surf. Coat. Technol.* **309** 445
[15] Ding Yu, Zhang Yu, Northwood D and Alpas A T 1997 *Surf. Coat. Technol.* **94** 483

SPH simulation of plastic deformation in high velocity impact welding process of 6061-T6 alloy plates

Yu Yu Emurlaeva[1], **N S Aleksandrova**[1] **and I A Bataev**[1]

[1] Department of Materials Science and Engineering, Novosibirsk State Technical University, 20 Prospekt K. Marksa, Novosibirsk

E-mail: emurlaeva@corp.nstu.ru

Abstract. In this study, the plastic deformation process of two plates of aluminum alloy 6061-T6 was simulated at different collision point velocities using the method of smoothed-particle hydrodynamics (SPH). The results of numerical simulation were compared with the results obtained in experimental and theoretical studies of other authors. The data obtained indicate that the SPH method reproduces the wavy interface between the plates, which was previously observed in numerous experiments. In addition, SPH simulation allows one to study in better detail the features of plastic deformation occurring near the interlayer boundaries, in particular at the vortex zones.

1. Introduction

The formation of a joint during high velocity impact welding occurs due to a number of interrelated phenomena, including the cleaning of the impact surfaces by a jet in front of the contact point and the mutual plastic material flow in the contact area of the colliding plates. During the solid state welding process, the value and distribution of plastic deformation play a key role, which determine the formation of physical contact and bonding of metals. A large number of studies were aimed at creating and improving methods that provide qualitative and quantitative information about the patterns of plastic flow of material in the immediate vicinity of the weld line, since all processes responsible for the activation of surface layers occur in relatively small volumes of metal adjacent to the zone of physical contact.

All experimental methods for the analysis of plastic deformation can be divided into microstructural, in which the residual plastic strain is evaluated based on the change in the shape of natural markers such like individual grains or twins, as well as based on the use of artificially prepared layered inserts or monolithic inserts with special grids. The method of layered model inserts is based on fabrication of composite packages of thin foil material, which is close in mechanical properties to the studied one [1]. These packages are installed in preliminary prepared grooves in welded plates. After the welding process, the photographs of the packages are combined and the resulting image is mathematically processed. This method has several limitations and inaccuracies, it also requires complex and time consuming sample preparation. The method of mechanically prepared grid is as follows: the insert sample is cut from the center of the flyer or the base plate, the grid of the grooves is applied on its side faces, after that it is placed on its previous position. After the welding the insert is removed from the plate again and the materials strain is measured by the distortions of the grid [2]. However, this method is not suitable for the analysis of deformation with high collision point velocity since the grid is erased due to intensive flow of material in the vortex zones.

Currently, various methods of numerical simulation are used to analyze the processes of plastic deformation, heating and cooling of the material during explosion welding. By other authors, Euler, Lagrange, Euler–Lagrange grid numerical methods and their modified versions [3, 4] as well as gridless methods like smoothed particle hydrodynamics (SPH) [5 - 9] were used to simulate the high velocity impact welding process. The use of grid methods doesn't allow one to completely reproduce such phenomena as wave and jet formation. Moreover, the computational process can be interrupted for large values of deformation. Therefore, in this paper, the gridless SPH method was chosen for the study, which is widely used for problems with fast moving boundaries and intensive plastic material flow.

2. Selected methodology

The Ansys Autodyn environment, the smoothed-particle hydrodynamics method, the Johnson-Cook strength model, and the modified Mie-Gruneisen equation of state were chosen to simulate the process of plastic deformation under conditions of high velocity collision. Aluminum alloy 6061T6 was chosen as studied material. It should be noted that several experimental investigations [10–12], including the classic work of Wittman [10], were devoted to the study of explosion welding of this material. The data obtained in the experiments of other authors were used for comparison with the results of simulation performed in this work.

Table 1. Parameters of 6061-T6 alloy used in numerical sumulation.

Johnson-Cook strength model		
Parameter	Value	Dimension
Shear Modulus	26	GPa
Yield Stress	0.324	GPa
Hardening Constant	0.114	GPa
Hardening Exponent	0.42	-
Strain Rate Constant	0.002	-
Thermal Softening Exponent	1.34	-
Melting Temperature	925	K
Ref. Strain Rate (/s)	1	-
modified Mie-Gruneisen equation of state		
Reference density	2.7	g/cm^3
Grüneisen coefficient	1.97	-
Parameter C1	5.35	m/ms
Parameter S1	1.34	-
Reference Temperature	293	K
Specific Heat	$8.85 \cdot 10^{-4}$	kJ/gK

At the initial moment of time, two plates with a thickness of 6 mm were located at an angle to each other; the upper plate had an initial impact velocity Vp, which defines a collision point velocity as:

$$V_C = \frac{V_P}{2\sin\frac{\alpha}{2}} \quad (1)$$

where α is collision angle.

Several collision regimes were simulated in this study. For all of the simulation the collision angle α was 25°. The following impact velocities were simulated (the corresponding values of V_C are given in brackets): 216 m/s, 563 m/s, 909 m/s, 1255 m/s and 1602 m/s (500 m/s, 1300 m/s, 2100 m/s, 2900 m/s and 3700 m/s).

The materials constants used for the simulation are shown in Table 1.

3. Results of numerical simulation

An effective (equivalent) plastic strain was chosen as one of the main parameters used in this work for the assessment of plastic deformation in the contact zone, which value is expressed as an integral of plastic deformation increment over a period of time t:

$$\varepsilon_{ef} = \int_0^t \sqrt{\frac{2}{3} d\varepsilon_{ij} d\varepsilon_{ij}} \quad (2)$$

There is an exponential increase in the effective plastic strain as it approaches the interface, which is well observed at low collision point velocities (Figure 1).

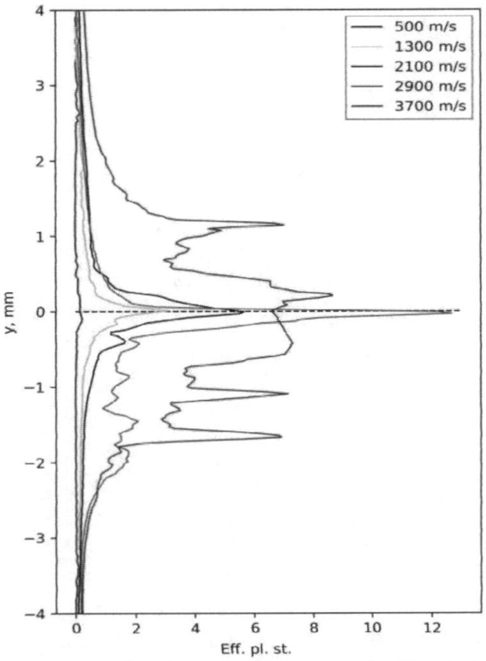

Figure 1. Effective plastic strain distribution for different values of V_C (dotted line corresponds to the position of the interface).

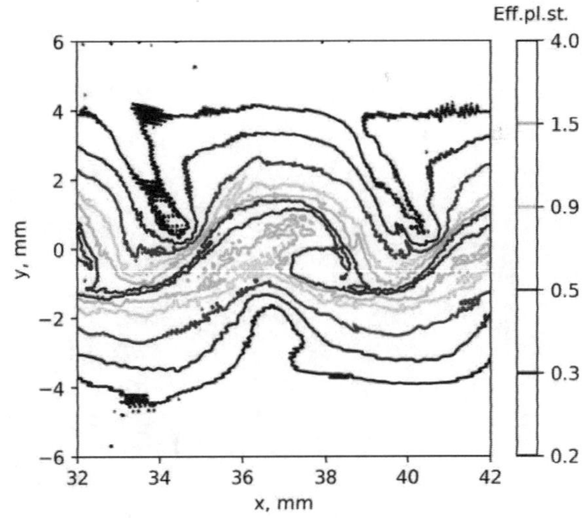

Figure 2. Equal strain zones (V_C = 2900 m/s, α = 25°).

IOP Conf. Series: Materials Science and Engineering **795** (2020) 012003 doi:10.1088/1757-899X/795/1/012003

Analysis of the resulting plots of effective plastic strain in the heat-affected zone allowed conditionally dividing the deformed metal layer into three characteristic zones: (i) the zone adjacent to the weld line, in which the largest strain values are achieved, or the zone of maximum material flow; (ii) the transition zone in which the deformations are distributed more uniformly; and (iii) the peripheral zone, the strain of which was close to zero. A schematic arrangement of equal strain zones for one of the studied regimes is presented in Figure 2. The obtained data are consistent with theoretical data and observations of the previous studies [1, 2].

The selected method of numerical simulation made it possible to study the nature of plastic deformation in the near contact region for both plates – the flyer one and the base one. It may be noticed that when two plates having similar composition are welded the distribution of strain near the interface is rather symmetric for most of the simulated regimes.

The solid-phase mechanism for the formation of a strong connection of metal layers during collision interaction involves the generation in the heat-affected zone of a certain level of shear deformations responsible for the activation of the contact surfaces and the development of processes of mutual plastic flow of the flyer and base plates. The biaxial stress-strain state is realized during the explosive welding of metals [2], a necessary condition for the formation of the joint is to achieve the maximum value of the critical strain. In the case of collision at relatively low velocities (collision point velocity $V_C = 500$ m/s, impact velocity $V_P = 216$ m/s and collision angle $\alpha = 25^{\circ}$), the strain in the contact zone didn't reach the threshold value to form a region of maximum material flow. This in good agreement with classical models [10], according to which such collision regime doesn't lead to the formation of a bond. It seems that a necessary condition for a successful explosive welding process is the presence of all three regions of plastic flow.

According to the obtained data, the process of jetting begins when the effective plastic strain near the

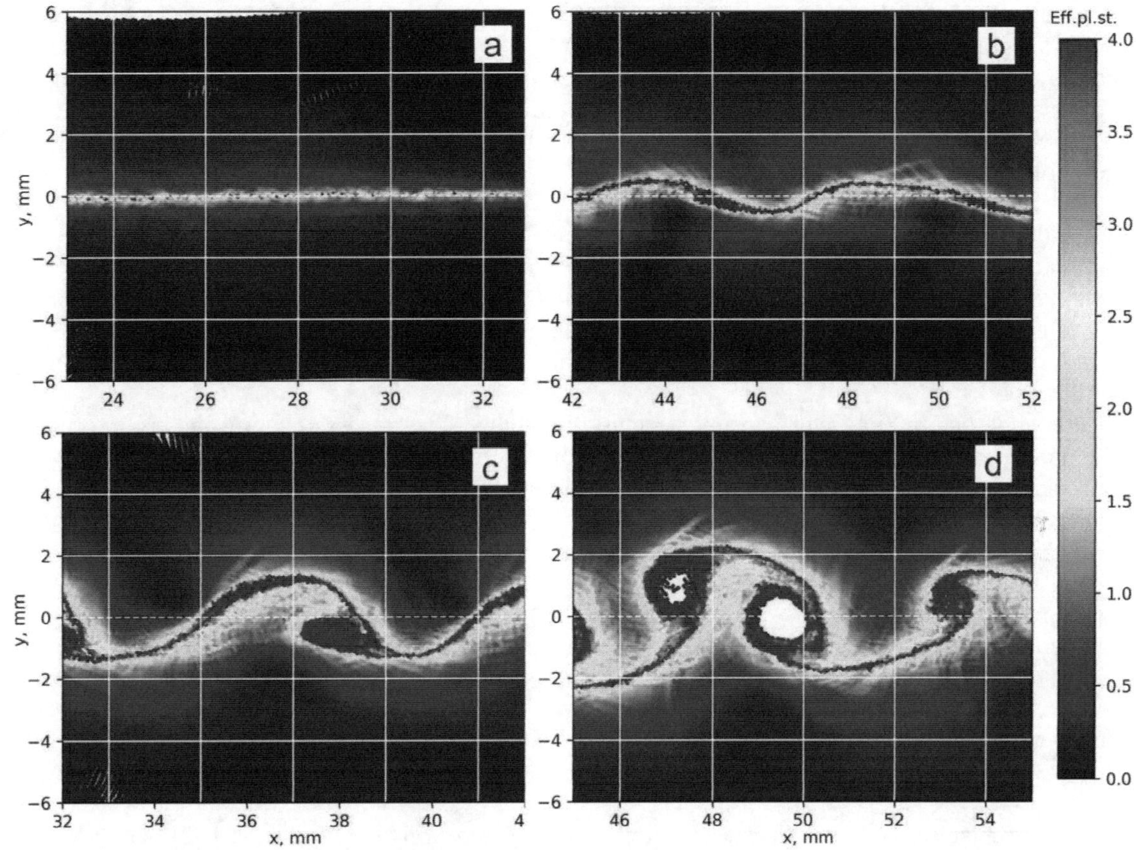

Figure 3. Distribution of the effective plastic strain in the plates interface: a - V_C = 1300 m/s; b - V_C = 2100 m/s; c – V_C = 2900 m/s; d - V_C = 3700 m/s.

interface reaches value of ε = 3 (Figure 3-a), which corresponds to the welding regime with V_P= 563 m/s, (V_C = 1300 m/s) and α = 25°. For this collision regime, the strain is predominantly localized in a narrow contact zone within a distance of 2 mm from the interface, while the largest values of strain are observed in a layer of 0.33 mm from the interface. The interface has a straight shape, thus the residual strain field along the interface is rather uniform.

The process of wave formation starts when the effective plastic strain reaches value ε ≈ 6, which corresponds to the welding regime V_P = 909 m/s (V_C = 2100 m/s) and α = 25°. The wave amplitude in this welding mode is about 1 mm (Figure 3-b), the effective plastic strain in near the wave area is in the range from ε =1 to ε = 6. The dependence of strain on the distance from the interface has an exponential form, which is in a good agreement with the experimental results obtained by Chugunov et al. [1].

Chugunov et. al. [1] also reported that experimental estimation of shear strain using inserts with preliminary prepared grids was difficult in cases of intense waving with formation of vortices, because the grid was erased, or it was difficult to analyze the samples using standard metallographic methods. The smoothed-particle hydrodynamics method made it possible to calculate the effective plastic strain even for cases with fast-moving collision point and a complex interface morphology. The obtained strain values for one of the sections are presented in Figure 1 (green line) at V_C = 2900 m/s, V_P = 1255 m/s, α = 25°. The dependence of the plastic strain on the distance from the interface also has an exponential form, however, the zone of plastically deformed material in the base plate is significantly larger than in the flyer plate. This fact requires special attention in the production of composite bimetals, for which residual strain has a key role in the exploitation.

The welding mode at the collision point velocity $V_C = 3700$ m/s with the collision angle $\alpha = 25^\circ$ is in the region of optimal collision regimes according to the Deribas hydrodynamic model [13] and is close to the upper welding limit according to the Wittman model [10]. The corresponding curve (Figure 1) shows a significant plastic flow of material at a depth of more than 3 mm. The curve also differs from those for the previously described regimes. The maximum value of plastic strain is lower compared to the previous regimes having lower collision point velocity, however, the thickness of the deformed area increases, which is associated with an increase in wave amplitude, as well as with a change in their morphology (Figure 3-d).

4. Conclusions

Summarizing the obtained results, we can conclude that the plastic flow of the metal in the heat-affected zone has a decisive role in the formation of a solid-phase bonding in high velocity collisions. It is subjected to strict regularity depending on the welding parameters, providing an opportunity to manage purposefully the activation processes of the contacting surfaces and control the properties of the obtained bond.

The smoothed-particle hydrodynamics method reproduces the process of jetting and wave formation. The data obtained using numerical simulation with different welding regimes are in good agreement with the experimental data presented in the experimental studies of other authors. It is possible to use this approach for further study of welded plates of dissimilar materials with various physical and mechanical properties.

Acknowledgments
This work was supported by Ministry of Education and Science of the Russian Federation according to Federal Task 11.7662.2017/BCh.

References

[1] Chugunov E, Kuzmin S, Lysak V and Peev A 2001 *Fizika i khimiya obrabotki materialov* **3** 39–44
[2] Kriventsov A and Sedykh V 1969 *Fizika i khimiya obrabotki materialov* **1** 132–141
[3] Nassiri A, Chini G, Vivek A, Daehn G and Kinsey B 2015 Arbitrary Lagrangian–Eulerian finite element simulation andexperimental investigation of wavy interfacial morphology during high velocity impact welding *Materials and Design* **88** 345–358
[4] Tabbataee M and Mahmoudi A *Finite element simulation for explosive welding* (IST Dep., HST School, MDH University)
[5] Bataev I, Ivanov I, Malyutina Yu, Émurlaev K and Émurlaeva Yu 2019 High plastic deformations and high cooling rates at the interface of explosion-welded materials *Metal Science and Heat Treatment* **60** 674-679
[6] Monaghan J 2012 Smoothed particle hydrodynamics and its diverse applications *Annu. Rev. Fluid Mech.* **44** 323–346
[7] Wua Y, Lub J, Tanb S, Jianga F and Suna J 2018 Modified implementation strategy in explosive welding for joining between precipitate-hardened alloys *Journal of Manufacturing Processes* **36** 417-425
[8] Zhanga Z, Fengb D and Liua M 2018 Investigation of explosive welding through whole process modeling using a density adaptive SPH method *Journal of Manufacturing Processes* **35** 169-189
[9] Bataev I 2017 *Metal Working and Material Science* **4** 55–67
[10] Wittman R 1973 The influence of collision parameters on the strength and microstructure of an explosion welded aluminum alloy *Proceedings of the Second International Symposium on the Use of Explosive Energy in Manufacturing* (Marianskie Lazni, Czechia) p 153–168

IOP Conf. Series: Materials Science and Engineering **795** (2020) 012003 doi:10.1088/1757-899X/795/1/012003

[11] Grignon F, Benson D, Vecchio K, and Meyers M 2004 Explosive welding of aluminum to aluminum: analysis, computations and experiments *International Journal of Impact Engineering* **30** 1333–1351
[12] Zhang Y, Babu S, Zhang P, Kenik E and Daehn G 2008 Microstructure characterisation of magnetic pulse welded AA6061T6 by electron backscattered diffraction *Science and Technology of Welding and Joining* **13** 467-471
[13] Deribas A 1980 *Physics of Hardening and Explosion Welding = Fizika Uprochneniya i Svarki Vzryvom* (Novostbirsk: Nauka)

IOP Conf. Series: Materials Science and Engineering **795** (2020) 012004 doi:10.1088/1757-899X/795/1/012004

Main problems of mathematical modeling high energies plasma technologies

V I Bogdanovich and M G Giorbelidze
Samara National Research University, 34, Moskovskoe Highway, Samara, 443086, Russian Federation

E-mail: bogdanovich@ssau.ru; m.giorbelidze@ssau.ru

Abstract. Some fundamental problems of mathematical modeling of processes to get coatings using vacuum ion-plasma method have been formulated. It has been introduce a notion of an active surface zone to describe non-thermal activation. It has been shown that non-thermal activation processes can be explained because of the surface focuson effect. Main evolution stages of conversion of absorbed particles have been described based on probability theory. Some general theoretical relations applied in the theory of formation of a new phase on a surface of a solid body have been defined.

1. Introduction

Plasma technology of high energies (PTHE) for obtaining special coatings in a vacuum of metals, their compositions, and ultra-hard compounds are in ever-growing use to increase operational life and reliability of various engineering products [1, 3-11].

The essence of this technology is the following. Using an electric-arc evaporator or accelerator a plasma jet is developed in a vacuum arc of a cathode form. This plasma jet consists of ions and atoms of a cathode's metal (or cathode's metals in a multi arc mode) and electrons. Main features of this plasma jet are high degree of plasma ionization (more than 0.5), high initial ion energy (more than some tens of eV), average multiplicity of ion ionization exceeding unity, and high density of a metal mass flow in a plasma jet (corresponding to a coating growth rate 0.1 .... 1 mcm/min) [1, 2].

After various treatment, this plasma jet is directed to the condensing surface that is under adjustable negative potential 0…2 kV. As a result, ions get additional kinetic energy ezU_c in a Debye layer near the part surface, then after neutralization, they get to a condensing surface. Depending on the part potential value U_c, coating formation, ion sputtering of the surface or even ion implantation can happen. While supplying reactant gases into a vacuum chamber it is possible to get a coating of various compounds (nitrides, carbide, oxides, etc.).

Over the past period of development of this technology some fundamental and applied tasks have been stated and solved, they have also allowed to provide its wide introduction into an industry. However, prospects for further development of PTHE as well as other technologies for coatings applying ion-bombardment of a surface are connected with the necessity of solving a range of fundamental, applied and technical problems.

2. Main processes typical for ion-plasma technologies

Mathematical modeling of plasma technologies of high energies (PTHE) is to describe the following main processes typical for ion-plasma technologies:

1. Interaction of a plasma jet with a 'cold' gas while transporting the stream from a cathode to a condensing surface.
2. Electrical exchange of moving non-isothermic quasi-neutral plasma with a solid body surface.
3. Interaction of atomic particles of average energies 10…1000 eV with a solid body surface. Influence of these particles and energies on reaction rate constants of absorption, hemisorption, surface migration, desorption, dissociation and embedding into active growth center of coatings.
4. Influence of technological heredity (initial and acquired during the coating growth) on product properties with coatings, and control of these properties.
5. Optimization of structure complexity to obtain extreme operating properties of multicomponent systems.
6. Influence of basic controlling physical parameters of deposition on physicochemical properties of coatings and engineering-and-economical performances of the deposition process.
7. Rationalization of practical fields of application of multicomponent and multilayer coatings.

In addition to these tasks specific for ion-plasma deposition it is also necessary to solve both general tasks connected with thermalphysic processes and mechanics of deformable bodies with variable geometry [1, 3].

Essential feature of physical processes we are to consider within the framework of PTHE is a range of kinetic energies of plasma's atomic particles. This range is 1…200 eV under transportation of a plasma stream, 1…500 eV in case of coating condensation, and 1…2000 eV under ion deposition. Modern known fundamental results of description of atomic particles and molecules interaction have been basically obtained for the energies deliberately lower 1…10 eV and deliberately higher 10^3 eV. As a result, one of the most important problems arising in considered energy ranges is a reasonable choice of the potential of interatomic interactions of heavy particles.

Conducted analysis of classical potentials (Born-Mayer, Mare, Bohr, power, etc.) as well as potentials obtained on the basis of statistical methods taking into account electron interaction (Thomas – Fermi – Firsov, Hartree – Fock – Sletter, free electronic gas, etc.) has shown that in the considered energy regions values can differ by an order and more [2]. Absence of reliable experimental data on scattering cross-sections in the energy region 1…500 eV does not allow to take clear choice of the interaction potential, and consequently to base physical process with needed definiteness. Widely used in papers hard spheres approximation could not be reasonable because it does not allow describing the known experimental fact of prevailing scattering at little angles within this energy range. The only excuse for using this approximation in cascades of shifts of a solid body is rather weak dependence of cascade average characteristics on the type of equation for the energy transfer cross-section.

The second essential feature of the considered technology is the fact that coating formation and heterogeneous reactions take place under intensive bombardment of a surface by fast atomic particles with energies essentially higher than thermal energies, but with essentially lower energies considered for calculation of cascades of shifts. Consequently, well-developed methods of statistical physics (thermal interaction region) and method of kinetic equations (region of ion implantation and scattering) cannot be applied without certain clarifications. Therefore, a necessity arises to develop new approaches to description of processes on solid body surfaces when forming coatings under ion bombardment.

One of possible variants of description such processes was suggested by the authors in [4].

The essence of it that the driving force of some processes on the surface is considered as a superposition of impacts based on thermal and non-thermal activation. Thermal and activation is determined by classical equilibrium distributions on energies, and it results in known equations of theory of transition state Eyring-Evans-Polanyi or in formulae of Ya.I. Frenkel type with Boltzmann multiplier.

3. Effect of formation of surface focusons

Non-thermal activation of these processes is stipulated by an ion bombardment of a surface and kinetic energy transfer from fast particles. This kinetic energy is transferred to atoms of a crystal lattice during sequential collisions. It also develops active area inside and on a solid body surface. For the lifetime of this active area, various physical-chemical processes with non-thermal activation by phonons, electrons and photons take place. Without discussing the nature of this activation, we can state that non-thermal activation of a surface process of the given type takes place in some active area S_A, and this area appears under high-energy particle that have impacted the surface. To be more specific, it is assumed that the size of an active area S_A such that probability of the given type of surface processes is close to unity. Naturally, at a first approximation, size of this area is proportional to the released energy W_i and inversely proportional to the activation energy U_A of the considered physical-chemical process. Thus, area value of an active zone can be introduced at a first approximation as:

$$S_A = A\frac{W_i}{U_A}\cdot\frac{1}{n_r}, \tag{1}$$

where n_r – reticular density of surface atoms, coefficient A has a physical meaning of a number of atoms on the surface available for sorption and having obtained energy U_A sufficient for the activation of physical-chemical processes of the given type.

For example, using introduced notion of a magnitude of an active zone S_A we determine a reaction constant of non-thermal dissociation of nitrogen molecules adsorbed on a surface with the density n. For some time period dt a number of high-speed atoms of metal $N = j_i S dt$ with stream density j_i fall on the surface with area S. Each of these atoms develops active zone with the area S_A. According to (1) it provides dissociation $n \cdot N \cdot S_A$ of absorbed nitrogen molecules. Using a definition of a macroscopic reaction rate constant λ_H of non-thermal dissociation of nitrogen, we get;

$$\lambda_H = A\frac{W_i}{U_A}\cdot\tilde{j}_i, \tag{2}$$

where $\tilde{j}_i = j_i / n_r$ - reduced stream density; U_A – in the given case the energy of activation of molecule dissociation n_r in absorption state on a surface.

This phenomenological approach was experimentally proved when obtaining coatings of nitrides [4] in the region of low temperatures, but it resulted in necessity of substantiation of physical mechanism of such dissociation. First, physical meaning of a magnitude S_A – is actually a middle cross-section of interaction of a fast atom with a solid body surface. But obtained experimental magnitude S_A is evaluated in area occupied from some dozens to some hundreds of surface solid body atoms. Secondly, activation energy of desorption N_2 is of 0.4…..0.8 eV, but the bond energy of nitrogen atoms in a molecule, even taking into account weakening of this bond by the surface interaction, cannot be less than 6…8 eV. Therefore, mechanism of kinetic energy transfer from fast atomic particle to the atoms of active zone and the mechanism of energy transfer from the atoms of active zone to the absorbed molecules of nitrogen is to be such that the possibility of strong bond breakage of atoms in a nitrogen molecule without kinetic energy transfer to the whole molecule (in the direction perpendicular to the surface) would appear, and it would be sufficient for its desorption.

To answer the raised questions the authors have suggested the mechanism of occurrence and development of surface focusons of Silsby type [5]. Really, when the energy of impacting particles is less than 300…500eV, their scattering cross-section on solid body atoms is close to a half of a crystal lattice constant as early as during the first interactions. In this case, focusons develop at certain angles of scattering directions along surface crystallographic axes of a crystal lattice. These focusons provide energy and impulse transfer at very long range (in comparison with interatomic ones) due to consecutive elastic collisions of atoms of the same mass.

In case of such consecutive collisions of surface atoms of a solid body, absorbed molecules and atoms in the potential field of these surface atoms also get certain impulse and kinetic energy along the surface direction of a solid body. If an absorbed atom gets such impulse and energy, it passes to another sorption center. If such an impulse and energy is obtained by the atoms not simultaneously or by one of atoms of an absorbed molecule of nitrogen, bond breakage N-N or excitation of violent oscillations happens along this bond weakened by the surface interaction with subsequent dissociation of a molecule.

Consequently, in both considered cases we deal with non-thermal activation of surface migration processes and with molecule dissociation on a surface. Besides, in contrast to thermal activation, described processes slightly depend on the surface temperature, and absence of impulse component perpendicular to a solid body surface eliminates the problem of keeping particles in absorbed state after obtaining rather high kinetic energy from a focuson.

Quantitative description of influence of effect of surface focusons initiation on coating formation processes, chemisorption, etc. can be done using introduced notion of an active zone of non-thermal activation on a solid body surface [2].

We assume that a fast atom with the energy W_i that has fallen on the surface is able to develop kW_i/W_f - surface focusons (k – efficiency coefficient of surface focusons production with the average energy of their production W_f). Designating as $\lhd W$ average losses of focuson's energy in one act of sequential collisions in a chain (taking into account transfer of part of energy to absorbed atoms and molecules) we get that each of the focusons embraces $N = W_f / \lhd W$ surface atoms. Magnitude N can be introduced as $N = S_{\Phi,1} \cdot n_r$ where $S_{\Phi 1}$ - effective surface zone area of non-thermal activation produced by a focuson.

Thus, total effective area S_A of an active zone of non-thermal activation is:

$$S_A = \frac{kW_i}{W_f} \cdot S_{\Phi,1} = \frac{kW_i}{\lhd W} \cdot \frac{1}{n_r}. \tag{3}$$

Introducing in (3) the magnitude A based on relation $A = kU_A / \lhd W$, we can form (3) similar with relation (1) obtained phenomenologically.

Considered effect of arising of surface focusons is at the initial stage of development, and it requires detailed theoretical and experimental research because of high significance of ion-plasma technologies for practical purposes.

4. Calculation of constants of surface reaction rate in case of plasmochemical synthesis of coatings

For logical completion of analysis of influence of thermal and non-thermal effects on heterogeneous processes, it is necessary to develop a calculation algorithm of rate constants of corresponding surface reactions. In this connection, we consider some features of this calculation with respect to an absorbed molecule N_2, taking part in formation of a nitride coating.

Molecule N_2 having undergone thermal accommodation falls on a certain sorption center. This molecule leaves this sorption center in a certain time because of thermal or non-thermal activation. With some probability, it can fall on another sorption center (surface migration) in a similar molecule state or it can desorb from the surface or dissociate into atoms. If a molecule has migrated to the second sorption center, from this center as well as from the first one it can take part in three mentioned above outcomes with a certain probability. Thus, determination of desorption and dissociation rate constant is a probabilistic analysis of molecule behavior at two structural levels. The first level is a possible particle transformation at leaving from this sorption center. The second level is a possible particle transformation if it migrates to different sorption centers.

First, let us consider probabilistic description of possible particle transformations at the first structural level. Suppose in time t=0 a particle falls on some sorption center, and at some random time

IOP Conf. Series: Materials Science and Engineering **795** (2020) 012004 doi:10.1088/1757-899X/795/1/012004

$t=T$ it can leave the center with three possible outcomes. The first outcome is a migration, the second one is desorption from the surface, and the third outcome is dissociation in atoms. Each of these outcomes can be realized with the intensity λ_i (i=1, 2, 3) in absence of others.

In case of purely thermal activation, intensity of each outcome can be determined using Ya.I. Frenkel formula or using formulae of transition-state theory:

$$\lambda_{i,T} = \nu \exp(-U_i/kT), \tag{4}$$

where U_i – activation energy of an appropriate process. To simplify, preexponential factor is taken as ν - a frequency of a particle oscillation in a bound state.

In case of atom bombardment of the surface by fast atoms these outcomes for an absorbed molecule N_2 happen because of non-thermal activation of an appropriate process mentioned above. Therefore, for the intensity of non-thermal activation of desorption happening due to absorbed molecules sputtering we get [4]:

$$\lambda_{2,H} = \alpha_{P,2}\tilde{j}_i, \tag{5}$$

where $\alpha_{P,2}$ – sputtering ratio of absorbed molecules.

Intensity of non-thermal dissociation is determined by (2), and we assume $\lambda_{3,H} \equiv \lambda_H$.

Thus, having considered each of three possible outcomes separately we can state that the given outcome is realized with intensity $\lambda_i = \lambda_{i,H} + \lambda_{i,T}$. These events happen because of accidental, independent, and individual for each event reasons.

Statistical description of a probability of each outcome can be the following. A particle is on the given sorption center for some time $[0,t]$. It can leave it with intensity λ_i at some accidental time T. Set $\{T\}$ when considering statistically equivalent experiments is a set of random events. We break up $[0,t]$ into N equal parts in length t/N. Then possibility of leaving of a particle in any of intervals t/N is $\lambda_i t/N$, and possibility of an opposite event over the whole interval $[0,t]$ is determined as a product probability of this opposite event $(1-\lambda_i t/N)$ at all N intervals:

$$P\{T > t\} = \lim_{N\to\infty}(1-\lambda_i t/N)^N = e^{-\lambda_i t}. \tag{6}$$

Hence, probability that a particle leaves the given sorption center on i-reason for the time t is:

$$P\{T < t\} = 1 - e^{-\lambda_i t}. \tag{7}$$

Consequently, for individual i-event we have a demonstrative probability distribution law (7), and its distribution density is:

$$f_i(t) = \lambda_i e^{-\lambda_i t}. \tag{8}$$

When considering three independent possible outcomes, possibility to discover a particle in the given center for some time that is greater or equal t equals possibility of failure of these three events:

$$P\{T_1 > t, T_2 > t, T_3 > t\} = P_1\{T_1 > t\}P_2\{T_2 > t\}P_3\{T_3 > t\} = e^{-\lambda t}, \tag{9}$$

where $\lambda = \lambda_1 + \lambda_2 + \lambda_3$.

It follows that the mean lifetime of a particle on this center (mathematical expectation) is $\tau = 1/\lambda$.

Possibility that the outcome will be the first event (migration) is determined as a possibility of occurrence of the first event earlier then the second and the third ones:

$$P_1 = P\{T_1 < T_2, T_1 < T_3\} = \int_0^\infty f_1(t_1)\left[\int_{t_1}^\infty f_2(t_2)dt_2 \cdot \int_{t_1}^\infty f_3(t_3)dt_3\right]dt_1 = \frac{\lambda_1}{\lambda} \tag{10}$$

Occurrence of the second and the third outcomes is determined similarly:

$$P_2 = \frac{\lambda_2}{\lambda}, \quad P_3 = \frac{\lambda_3}{\lambda}.$$

Consequently, possibility that a particle out of sorption state will move to the adjoining center

(i=1), desorb (i=2) or dissociate (i=3) will be:

$$P_i(t) = P_i \cdot \left(1 - P\{T_1 > t, T_2 > t, T_3 > t\}\right) = \frac{\lambda_i}{\lambda}\left(1 - e^{-\lambda t}\right), \tag{11}$$

and appropriate probability densities will be

$$f_i(t) = \lambda_i e^{-\lambda t}. \tag{12}$$

Process rates (reaction rate constants k_i^*) of migration, desorption, and dissociation when the particle is on the same center is determined as mathematical expectations from corresponding intensity distributions:

$$k_i^* = \int_0^\infty \frac{f_i(t)}{P_i\{T_1 > t, T_2 > t, T_3 > t\}} f_i(t)dt = \frac{\lambda_i^2}{\lambda} = \lambda_i P_i. \tag{13}$$

Now let us analyze changes of reaction rate constants of desorption and dissociation taking into account migration of particles over desorption centers.

From each desorption center with the similar bond energy a particle is able to move to the adjoining center with possibility P_1, P_2, P_3 (10) and with intensity k_1^*, k_2^* and k_3^* (13) to move to the adjoining center, to desorb or to dissociate. There is no need to take into account influence of degree of filling of a surface on these processes when forming coatings in contrast to chemisorption.

Then possibilities of these outcomes from the first sorption center are P_1, P_2, P_3, from the second one possibilities of these outcomes are $P_{1,2} = P_1^2$, $P_{2,2} = P_1 P_2$, $P_{3,2} = P_1 P_3$, and from some j-center are $P_{1,j} = P_1^j$, $P_{2,j} = P_1^{j-1} P_2$ and $P_{3,j} = P_1^{j-1} P_3$.

Microscopic rate constants of desorption and dissociation are determined as mathematical expectations of these events:

$$\begin{aligned} k_2 &= \sum_{j=1}^{\infty} k_{2,j}^* \cdot P_{2,j} = k_2^* \sum_{j=1}^{\infty} P_{2,j} = \frac{\lambda_2^2}{\lambda} \cdot \frac{\lambda_2}{\lambda_2 + \lambda_3}, \\ k_3 &= \sum_{j=1}^{\infty} k_{3,j}^* \cdot P_{3,j} = k_3^* \sum_{j=1}^{\infty} P_{3,j} = \frac{\lambda_3^2}{\lambda} \cdot \frac{\lambda_3}{\lambda_2 + \lambda_3}. \end{aligned} \tag{14}$$

Relations (14) obtained on the basis of strict probabilistic analysis differ from similar relations used in adsorption and hemisorption theories [6] where it is assumed without proper basis that $k_i \sim \lambda_{i,T}$. It is justified because in these theories relations including the combination k_2/k_3, which is proportional (in case of purely thermal activation of the process) to the relation $\exp(-(u_2 - u_3)/kT)$ are researched experimentally. Besides, physical picture of the process does not undergo qualitative changes. But, when obtaining coating in conditions of combined effect of thermal and non-thermal activation, qualitative and quantitative physical picture essentially depends on the k_i kind [2].

5. Results and discussion

Result analysis of numerical simulation of various variants of kinetics of plasmochemical synthesis of nitride coatings [2] has shown that kinetic schemes with the same energy adsorptive state of molecular nitrogen have the same disadvantage that arises in mathematical modelling of most chemisorption processes. Really, classic calculations with one prechemisorption molecule state [6] due to the quick change of k_2 under temperature rise allow to describe adequately kinetics of these processes only when $T \leq T_{kp}$, where T_{kp} depends on a molecule bond energy in this only prechemisorption state.

For chemisorption of gases with purely thermal activation, such considering often gives satisfactory results because the small temperature interval around active chemisorption temperature is of practical interest. For example, in ion-plasma deposition formation of nitride coatings can be obtained within the range from 300 to 700…800 K. Therefore, development of an adequate mathematical description in a wide temperature range is of practical interest.

Authors [2] have determined that this problem is eliminated if we take into account possibility of molecule adsorption and possibility of changing of energy magnitude in case of molecule migration on a surface. This also allows to describe adequately kinetics of the process in a wide temperature range. In this case, multiplicity of prechemisorption molecule states (polysorption) and possible transitions between these states can be presented in the following scheme (Figure 1).

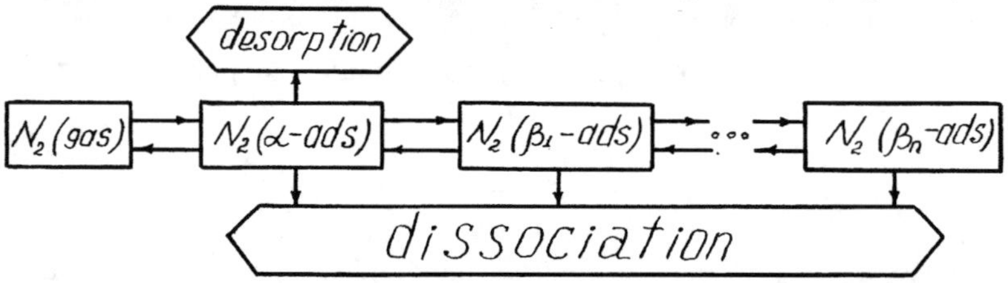

Figure 1. Transition scheme of an adsorbed nitrogen molecule between various energy states.

It may be shown that for the considered scheme (Figure 1), if the surface is slightly filled, the parameter $y=k_3/k_2$ may be the following:

$$y=(z(\alpha))^3+(\sigma_1')^2\left(\frac{\lambda_1(\alpha)}{\lambda_2(\alpha)}\right)^3\Phi(\beta_1,\beta_{2,}\ldots\beta_n), \tag{15}$$

$$\Phi(\beta_1,\beta_2,\ldots,\beta_n)=1-\gamma_2^2\left\{z(\beta_1)^3+\gamma_2^2+\tilde{\sigma}_2^2\Phi(\beta_2,\ldots,\beta_n)\right\}^{-1},$$

$$\Phi(\beta_i,\ldots,\beta_n)=1-\gamma_{i+1}^2\left\{z(\beta_i)^3+\gamma_{i+1}^2+\tilde{\sigma}_{i+1}^2\Phi(\beta_{i+1},\ldots,\beta_n)\right\}^{-1},$$

$$\Phi(\beta_n)=1-\sigma_{n+1}^2\left\{z(\beta_n)^3+\sigma_{n+1}^2\right\}^{-1},$$

$$z(\alpha)=\lambda_3(\alpha)/\lambda_2(\alpha),\ z(l)=\lambda_3(l)/\lambda_1(l),$$

$$\gamma_i=(1-\sigma_i)\sigma_i',\ \tilde{\sigma}_i=\sigma_i\sigma_i',\ l=\beta_1,\beta_2,\ldots,\beta_n,\ i=2,\ldots,n.$$

Research has shown that numerical simulation using (15) gives satisfying results when describing kinetics of plasma-chemical formation of nitrides on solid body surfaces for a wide temperature and energy range of incoming particles.

6. Conclusion

Thus, results of this work have allowed to formulate a set of urgent fundamental problems of obtaining coatings under ion bombardment and to outline approaches to their solutions. Moreover, some general relations applied in a theory of a new phase formation on a solid body surface have been specified.

Acknowledgments

The reported study was funded by the Ministry of Science and Higher Education of the Russian Federation.

References

[1] Barvinok V A 2005 *Plasma in Technology: Reliability and Resource* (Moscow: Nauka i technologii Publishers)
[2] Barvinok V A, Bogdanovich V I 1999 *Physical basis and mathematical modeling of vacuum ion-plasma deposition* (Moscow: Nauka i technologii Publishers)
[3] Bogdanovich V I 1999 *Engineering and Automation Problems* **2** 87-94
[4] Bogdanovich V I 1999 *Engineering and Automation Problems* **1** 65-71
[5] Oksengendler B L, Turaeva N N, Maksimov S E, Djurabekova F G 2010 Journal of Experimental and Theoretical Physics **111(3)** 415-420

[6] Roberts M W, McKee C S 1979 *Chemistry of the Metal-Gas Interface* (Oxford University Press)
[7] Gumen O, Bilyk I, Kruzhkova M 2020 *Lecture Notes in Civil Engineering* **47** 103-110
[8] Bobzin K, Brögelmann T, Kalscheuer C, Welters M 2019 *Thin Solid Films* **690** 137529
[9] Chen Z, Qiao L, Hillairet J, Song Y 2010 *Applied Surface Science* **492** 540-549
[10] Chang Y-Y, Cai M-C 2019 *Surface and Coatings Technology* **374** 1120-1127
[11] Potoczna-Petru D, Kępiński L, Krajczyk L 2005 *Materials Chemistry and Physics* **92(2-3)** 613-615

IOP Conf. Series: Materials Science and Engineering **795** (2020) 012005 doi:10.1088/1757-899X/795/1/012005

Modelling the interaction of stations in the port railway junctions

E Borodina[1], E Prokofieva[1], S Vakulenko[1] and A Petrov[2]
[1]Institute of Management and Digital Technologies, Russian University of Transport (MIIT), 127994, Moscow, 9b9 Obrazcova Street
[2]JSC «Institute of Economy and Development of Transport» (JSC «IEDT»)
105066, Novorjazanskaja str., 24, Moscow, Russia

E-mail: eskolesnikova@mail.ru

Abstract. The problem statement, qualitative analysis and main methodological approaches to the interaction of stations for the distribution of sorting and transfer work in port railway junctions with objects of railway infrastructure and rolling stock of various types are presented. The search for a solution provides for finding the optimal variant of interaction on the basis of combinatorial-optimization analysis using economic criteria, technical and technological parameters of stations and a plan of the formation of non-route gears in the area of pre-port transport hubs.

1. Introduction

Rational development of infrastructure and technology of operation of railway and sea transport hubs is the most important basis for the stable operation of the transport system of any state. The first initiative for the development of transport hubs as transport and logistics centers in Europe appeared within 60-70 years of the last century in France, Italy and Spain. But the intensive development began later in Germany, when the systematic cooperation of seaports and Railways was established [1; 2]. The main role of transport hubs with seaports is to export and import cargo.

At present, much attention is paid to the technological and well-coordinated work of transport junctions in Russia, including a port railway junction and seaport, the creation of efficient multimodal transport systems in the Baltic (Ust-Luga port), Azov-Black Sea basins (Novorossiysk and Tuapse) and the Caspian basin of the Republic of Kazakhstan (Aktau port). At the initiative of JSCo «Russian Railways», principally new technological documents – Unified Integrated Technological Processes (UITP) – were developed and approved:

- Ust-Luga transport hub [3];
- Novorossiysk transport hub and the ports of the Taman peninsula [4];
- The customs code of Aktau international sea trade port (AISTP) [5].

The UITP regulates the technology of traffic management and all railway operations, ensuring a smooth interaction of the participants in the transportation process: a dispatch center for transportation management, a shunting operator, state control bodies, operators of sea terminals, sea freight carriers, port administration, and units of a railway junction.

The customs code defines the established customs and practices of the port of Aktau, due to the technology of work, the rules regulate the procedure and conditions for the entry and exit of ships in and out of port, the mooring of ships to berths, unloading, loading and maintenance of ships, the supply of trains, and the relationship of the port with clients.

Despite large capital investments in the development of railway-sea transport hubs, they still experience technical and technological difficulties due to the uneven loading of the main devices of port railway stations, district parks, station necks, approaches, loading and unloading and storage routes.

Of great importance for the smooth operation is the construction of a rational system of maintenance of port complexes by rail. Fluctuations of the jets and the intensity of incoming car flows to the junction determine the main reason for the non-optimal distribution of sorting, shunting and transfer work between railway stations in the junction, which leads to a failure to ensure continuity of transshipment operations, untimely supply of cars to the terminals, re-processing of cars and, consequently, increased downtime wagons and the time of their turnover in the port.

In addition, the non-rhythmic and uncoordinated supply of goods to the transshipment points arises due to the lack of a single transport conveyor, disunity of ownership and management systems of transport and other entities involved in the transport process.

2. Literature review and defining the problem

The problems of interaction and technology of operation of subdivisions of the port railway junction, which include pre-port sorting stations (PPS), port stations (PS), district parks (DP), connecting, anchorage, handling and storage routes of sea terminals (T), distribution of sorting, freight and train work are devoted to the work of many authors and researchers [8-19]. The above problems have an important theoretical and practical significance. Theoretical – because it is necessary to develop principles and models for rational mutual placement, development of basic railway devices and interconnected technology between them, as well as models for managing train and car flows in accordance with the rhythms of loading and unloading cargo fronts of sea terminals. Practical – it is essential to reduce the time of processing vehicles, apply a direct version of cargo handling, accelerate their delivery, reduce material and labor costs based on the coordinated work of the transport hub as a whole, and introduce new integrated services in ports.

In the conditions of growing freight traffics, going in the ports, it is necessary to find the new technical, technological, organizational decisions of nascent problems. In this context, the reliability of transport is a key concept in the technology of organization and management of train and car flows. The concept of reliability is extremely important in assessing the capacity of transport networks to provide continuity in operation. The reliability of operational work depends, among other things, on the permissible modes of traffic flows through the railway network and has a significant impact on the cost parameters [8]. The problem of effective management of railway freight traffic especially in conditions of multiple operators require rationalization of transport process. The report [9] is devoted to consideration a simple mathematical model for the calculus of capacity in two different cases: for a homogeneous utilization, meaning that all trains have the same running speeds and for heterogeneous utilization, when the passenger trains have the higher speed and hence, the higher priority over the freight trains.

In the thesis [10] for the first time the structure of zones of effective interaction of railway and sea transport in port transport hubs was set out and presented, functional tasks were set, criteria and methods of optimal flow control in each zone were proposed. A method for calculating the coordinated supply of flows to cargo terminals of the port, implemented in the simulation model, is developed. The logistics of the interaction process developed in [10] on the basis of the proposed approaches within the existing information environment and the real structure of transportation management in JSCo «Russian Railways» are used to improve the operation of the port transport junctions, and the principles and methods of optimal flow control can be used to organize coordinated cargo delivery large consumers. The technique for calculating rational technical and technological parameters of transport junctions using simulation modeling can be useful when justifying investments in the development of junctions.

The purpose of the paper [11] is investigation of interaction of two contiguous types of sea and railway transport by means of the method of factor-by-factor analysis, allowing to produce organization of carriage process by the logistic system of freight traffics management. Temporal characteristics affect the question of principal distinction of work of railway and sea transport again. The interval between vessels arrival in port is greater by several fold the interval between trains. If port processes the certain amount of cargo, the port station cannot manage with the growing freight traffic. As a result, large numbers of freight trains are accumulated on the station approaches with cargo, going in an address of port. Therefore it becomes obvious, that the production capacities of port and station must be developed in parallels. Otherwise both participants of the mixed transportations will bear losses.

The scientific article [12] is devoted to the rational development of the port railway junctions in the Far East, in which the questions of the «throwing» of trains on approaches to the port railway junctions are investigated because of the absence of settling paths in the railway subsystems of the ports. The wagons awaiting submission to the berths occupy the ways of the sorting parks. This is the reason for the inefficient use of rolling stock and increased operating costs. To determine the path development of the port stations, it is proposed to apply the current method for calculating the track development of freight stations, and the track development of the pre-port sorting stations should be determined based on the needs of each port area and a specialized berthing complex. Parks of tracks for the sagging of wagons awaiting transshipment under the direct option («warehouse on wheels»), as well as wagons long waiting for the application for the port, depending on the topographical conditions and belonging to these ways, can be placed in the composition of any element of the berthing railway complex.

Excess downtime of wagons on railway devices and berth fronts on the example of the St. Petersburg Commercial Sea Port are investigated in work [13]. The specified problem is proposed to be solved by improving the technology of operation of the railway station and loading and unloading berthing fronts. For this purpose, it is proposed to introduce the position of the second car inspector in the staff list in order to reduce the downtime of the cars from arrival to the loading for loading; rationally distribute the shunting work on the car supply and picking to the port on exhibition routes by the locomotive station, and arrange wagons for loading and unloading fronts to perform a port locomotive, which will reduce the downtime in anticipation of the supply and picking of wagons by an average of 4 hours; ensure unimpeded removal of finished trains from Avtovo and Novy Port stations; to introduce a system for centralized management of a fleet of empty wagons with the planning of destination points for empty wagons after unloading, which will reduce the downtime of empty wagons of private owners; improve interaction between railway stations and the port in terms of harmonizing the approach time of ships and wagons using contact schedules for handling ships and wagons.

The thesis [14] is devoted to the increase of the track development of the stations and the development of constantly increasing volumes of cargo flows processed in seaports. In the current situation, it may be necessary to remove the existing junction or pre-port sorting station (CSS or PPSS, respectively) outside the railway junction and urban development, which simultaneously will solve the problems of the development of these stations, and also reduce the negative impact of stations on the environment of cities and human health. The solution of this task depends mainly on the choice of the most effective mutual placement of the port facilities: the USS or the PPSS, as well as the port stations (PS) and the district parks (DP). At the same time, the railway infrastructure should ensure the development of the required train traffic in the conditions of growing cargo turnover of the seaport.

The tasks associated with increasing the capacity of the ferry crossing by replacing old ferries with more powerful and capacious ones, with the establishment of effective interaction between the ports of the Azov-Black Sea basin for increasing passenger and cargo transportation are explored in the article [15] «On the rational distribution of rail freight traffic between the Azov-Black Sea basin in the foreseeable future».

In work [16] proposals are formulated to improve the indicators of the use of rolling stock (laden and empty trip) of the operator company for the carriage of goods in international corridors in the port transportation and technological system. Using the methods of geometric and economic-geographic modeling and their modifications to differentiate the spheres of influence of the station of wagons owned by the operator, an algorithm and method for distributing the rolling stock from loading stations to the port stations have been developed with the aim of reducing the duration of the car turnover. A decrease in operating costs for the transportation of laden wagons at port stations was obtained, depending on the length of the route, which gave an economic effect for the aggregate of regional distribution stations from the presence of wagons owned by the operator companies on them.

The delivery time of goods and empty wagons is of great importance for operators and cargo owners in terms of capital invested in goods and subsequently affects its turnover rate [17]. This approach provides probabilistic estimates of the rolling stock use efficiency for different approaches to the organization of car traffic volumes, as well as establishes the effect for each of the participants in the transportation process.

Significant contribution to the study of issues of interaction between rail and sea transport, underlying the modern theory of designing railroad and transport junctions, technologies for their effective work were also made by foreign scientists: G. Biddle, A. Jackson, J.E. Connor, I. Pachil, L. Frendrish, K.J. Holland, K. Hol and others.

The greatest interaction of maritime and rail transport in transport hubs is in container shipping. Therefore, at present, the task of the key operators of the European container business is to invest in the development of railway services and infrastructure on the approaches to the ports. More than half of the total investments of Hamburger Hafen und Logistik AG (HHLA), the largest terminal operator in the port of Hamburg, are in the development projects of railway operations and domestic terminals.

Foreign investors consider the European rail freight market to be promising. However, the uneven loading of terminals when entering mega-container ships remains a painful problem for ports. Large ships create a huge load on the entire port complex of Hamburg, where almost half of the containers are imported and shipped by rail. According to the executive board of the HHLA, an average of 53 trains are in the port waiting for cargo when entering a mega-container ship at the port.

The quality, reliability and profitability of the transportation process in the transport hubs are directly dependent on the technological stability of the port railroad junctions, the main elements of which are passenger, sorting, freight stations, facilities and devices serving the port's sea terminals [18]. Depending on the conditions for the development of the port railway junction, two or more sorting stations, a freight station, may be included in it, and in small junctions – one integrated station that performs all operations. The interaction of marshalling yards at the railway junction, as regards the distribution of sorting work between them, was investigated in [20-23].

The railway station is a complex system consisting of a number of different elements, which interact closely one to another in the process of operation providing mutual influence. The determination of rational parameters of railway stations is a complex multi-variant and multi-criteria problem. Currently, the simulation method is the most effective for modeling the technological processes of the station and evaluating its quality. In addition, the efficiency of functioning a station is largely dependent on the level of operational management by dispatch personnel. Simulation models can serve as an effective tool for analysis and assessment of functioning of the stations, their technical-technological and economic parameters that would allow making rational decisions to increase the profits of transportation by rail [24]. In paper [25] the area of rational use of trains' breaking-up order model, formalized in the form of stochastic programming problem is determined. This model depends primarily by two factors – accuracy of traffic forecasting and size of car processing.

As the analysis showed, in the previous studies, various options for port servicing are proposed without taking into account the reciprocal links between the pre-port sorting stations (PPS), port stations (PS) and district parks (DP). The costs associated with the incoming car traffic flow are taken into account, the cost of transferring traffic, delays of trains for non-acceptance of the PPS and the PS, as well as additional costs for increasing the capacity of the latter are not taken into account. The existing methods of rational servicing of port complexes by railway transport do not make it possible to plan and consider all distribution options for sorting cars and forming gears and trains, depending on the power of the incoming traffic of cars of various accessories.

At present, with the widespread introduction of automated information and logistics systems, the creation of the Automated Control System-Port (ACS-Port), and the organization of a common information space for participants in multimodal transport, it is necessary to develop an integrated method for distributing work in the port railway junction, which, in conditions of irregularity and fluctuations of incoming traffic flows, with disproportionate capacity and loading of railway devices, would provide for the establishment of a rational interaction between the railway transport and seaport in view of technical equipment and adopted technology work, as well as the implementation of various operational adjustment measures in the rail unit complex.

3. Research of the parameters of the solution

If the port junction has a significant cargo turnover and a sorting station at a sufficiently large distance from it, then various options for servicing the port [19] by railways arise (Figure 1).

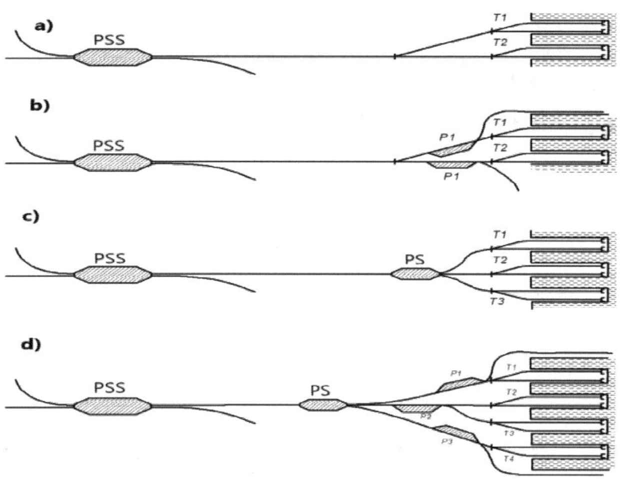

Figure 1. Schemes of relative location of railway equipment serving the port: a) servicing of sea terminals (T) by a pre-port sorting station (PSS); b) servicing of sea terminals for PSS and district parks (DP); c) servicing of sea terminals of the PSS and the port station (PS); d) service of sea terminals by all three railway devices

When deciding on the application of a scheme of the port service by rail transport, the condition of continuity of loading and unloading operations in the port and the arrival of wagons to each port area with goods of the corresponding categories should be taken into account first of all. The article [26] is devoted to the study of major port systems in Europe, in which a model of port classification was developed according to their actual position and regional role.

The operations for sorting wagons, their selection by separate areas and loading and unloading terminals of the port are performed at the pre-port sorting station. More detailed sorting – not only by cargo areas and berths of sea terminals, but also by ship holds at the port station, in special regional parks or directly on loading and unloading routes of sea terminals.

To solve this problem, the methods of analysis, applied combinatorics, simulation modeling of port stations and junctions are used. The original data of the problem: the operating car traffic volume having a discrete character on the approach to the port hub (the plan for the formation in the area of the pre-port railway hub on the basis of the reporting form DO-17), the layout of the main devices in the junction and their throughput, cost parameters.

4. Problem statement

The task is combinatorial-optimization, the nature of which is determined by the search of all possible options for the distribution of work, their technical and economic comparison and choosing the best option in the pre-port railway hub.

All operations carried out in the unit with wagons following to the port can be divided into three enlarged groups, then the operation of the port junction based on the UITP necessitates a significant modernization of the calculation dependences [6] in order to take into account modern features of the operation of the port units with the wagon flows of different cost estimates related to:

1) with the train flows entering the junction and leaving the junction – processing at the stations: $\mathbf{E_{pro}}$ – the costs of the owner of the public infrastructure and the railway carrier, as well as the owner of the non-public use path serving the terminal, and the shunting operator; accumulation at stations: $\mathbf{E_{accu}}$ – expenses of the operator of rolling stock; the formation of train trains and shunting transmissions at the stations: $\mathbf{E^{st}_{shun}}$ – the costs of the owner of the non-public way operating the terminal, and the shunting operator; train runs on sections, in- junction moves and connecting branches: $\mathbf{E_{runs}}$ – expenses of the owner of the general infrastructure and railway carrier; runs of single train locomotives due to traffic disparity in the sectors: $\mathbf{E_{sin.loco}}$ – expenses of the owner of the public infrastructure and railway carrier; downtime of wagons and train locomotives at stations: $\mathbf{E_{runs.wag}}$ – the costs of the operator of rolling stock, $\mathbf{E_{runs.loco}}$ – the costs of the owner of the general infrastructure and the railway carrier; delays of trains on unacceptable stations: $\mathbf{E_z}$ – expenses of the owner of an infrastructure of the general using and a railway carrier.

2) with the processing of intra-junction wagons – the organization of the transfer between the stations: $\mathbf{E_{trans}}$ – the cost of the owner of the public infrastructure and the railway carrier and shunting of the sea terminals: $\mathbf{E^{T}_{shun}}$ – loading and unloading operations in the port with the cars of the transfer train – the costs of the owner of the non-public way, and the shunting operator.

3) with the change of technical and technological parameters of stations, district parks and sea terminals of the junction in the redistribution of the car traffic volume and transformation of their parameters: $\mathbf{E_{tex}}$ – the costs of the owner of the public infrastructure and the railway carrier and the costs of the owner of the non-use route serving the terminal and the shunting operator.

Limitations of the task are determined by the stationary resources of the transport hub (approaches, in-junction hauls and connections, track development of stations, district parks, non-public routes, technical equipment of technological lines, the presence of a contact network and the type of current, fixed reloading and traction devices of terminals) and mobile resources and shunting locomotives, operational staff, mobile reloading and traction devices of terminals).

The task of justifying the technological parameters of the port railway junctions is combinatorial-optimization, the nature of which is determined by the nonlinearity and integer of a number of components of the objective function. In this case, the target function itself cannot be written down as a simple sum of the component costs associated with different train flows and with the technical and technological parameters of the components of the junction. As indicated above, the costs are borne by different participants in the transportation process, which requires decomposition into a series of single-objective tasks with subsequent coordination of the results.

The formal formulation of the problem is formulated as follows: it is necessary to choose such an option for organization of sorting in the conditions of current operation and for the future, in which the functioning of the railway port junction is ensured with minimal costs.

There is a managed system – a set of stations: pre-port sorting (PSS), port stations (PS) or district parks (DP), sea terminals with access roads of industrial stations (T) performing sorting work in the port junction. As control actions are variables: $\mathbf{K_r}$ – the number of assignments of the trains being formed on the interacting stations **r**; $\mathbf{N_{distr.r}}$ – car traffic volume between stations, involved in the distribution of sorting work; $\mathbf{n_{frei.x}}$ – number of freight trains between stations by section (in-unit connecting line) **x**. The state of the system is understood as the distribution of sorting work. Then the set of controls is represented as follows:

$$\mathbf{S_i = \{K_r;\ N_{distr.r};\ n_{frei.x}\}} \tag{1}$$

The choice of the optimal variant of the distribution of sorting work is connected with the choice of such control, in which the objective function reaches the minimum value,

$$\mathbf{E_{min} = min\{E\ (S_i)\}} \tag{2}$$

S

Target function – total average daily costs – includes components associated with various train flows and with technical and technological parameters of the stations. Finding a rational variant of the distribution of sorting work in the port junction reduces to solving the objective function:

$$\mathbf{E} = \sum \mathbf{E}_{pro} + \sum \mathbf{E}_{accu} + \sum \mathbf{E}_{shun} + \sum \mathbf{E}_{runs} + \sum \mathbf{E}_{gen.mov} + \sum \mathbf{E}_{runs.wag} + \sum \mathbf{E}_{runs.loco} + \sum \mathbf{E}_{z} + \sum \mathbf{E}_{trans} + \mathbf{E'}_{shun} \pm \sum \mathbf{E}_{rex} \rightarrow \mathbf{min} \quad (3)$$

under the following restrictions:

$$\mathbf{k}_r \le \mathbf{k}_{max.r} \quad (4)$$

$$\mathbf{N}_{obl.r} + \mathbf{N}_{distr.r} \le \mathbf{N}_{t.r} \quad (5)$$

$$\mathbf{N}_{frei.x} \le \gamma_{sec.x} \mathbf{n}_{cap.frei.x} \quad (6)$$

$$\mathbf{M}_{need.g} \le \mathbf{M}_{max.g} \quad (7)$$

$$\mathbf{N}_i \ge \mathbf{N}_{min.i}, \quad (8)$$

where $\mathbf{K_r}$ is the number of train assignments generated in the sorting system r;
$\mathbf{k_{max.r}}$ – the same as the maximum allowable for given processing sizes in the sorting system r;
$\mathbf{N_{obl.r.}}$, $\mathbf{N_{distr.r.}}$, $\mathbf{N_{t.r}}$ – car traffic volume processed in the sorting system r – mandatory, distributed and technically acceptable for a given number of appointments, wagons / days;
$\mathbf{n_{.cap.x}}$ – the capacity of the section (in-house course, connecting line) x for freight traffic, trains / days;
$\gamma_{sec.x}$ – permissible level of use of a throughput of a site (in-house course, connecting line) x;
$\mathbf{M_{need.g}}$, $\mathbf{M_{max.g}}$ – the required and maximum permissible fleet of train locomotives of the **g**-ing specialization (the treatment area);
$\mathbf{N_i}$, $\mathbf{N_{min.i}}$ – destination power **i** – respectively calculated and minimally permissible due to timely delivery of goods, wagons / days.

Restrictions (4) and (5) are interdependent. There is an overflow of sorting tracks under continuing condition of railway infrastructure and the growth of the incoming car traffic flow to the processing. As a result, it is necessary to allocate additional rail tracks, which reduces the possible number of generated appointments of the plan of formation. Figure 2 shows an example of this dependency, which demonstrates the dynamics of the power of the number of destinations from the allowable value of processing of the incoming car flow. At the same time, calculation of the operational performance is made on simulation models of port junctions and stations.

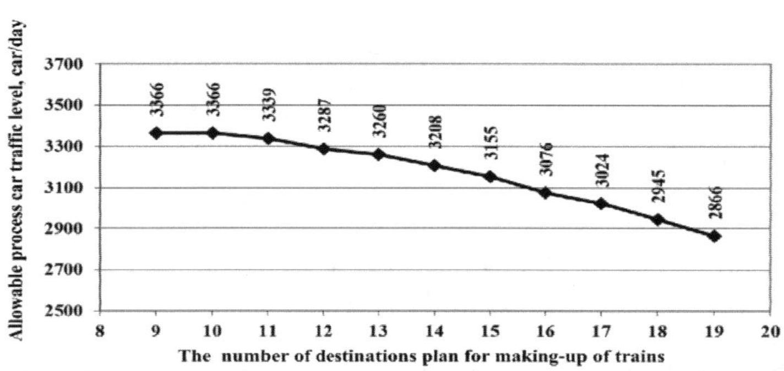

Figure 2. The graph of dependence of permissible processing of the car traffic flow on the number of generated appointments

The information structure of the studied task is such that the controlled variables should be determined on the average daily period of the plan for the formation of freight trains and the timetable when trains approach the port.

Costs are determined by the car- and train-hours of finding cars and trains in the junction, taking into account the technical and technological parameters of the stations and throughput capacities of the intra-junction lines.

The objective function (3) is written in a general form. At the same time, each participant of the transportation process (the owner of common infrastructure, the railway carrier, the owner of the railway line of uncommon use, the shunting operator and the operator of rolling stock) has its components of the cost in the objective function (3). The choice of a solution within the limits of constraints (4 - 8) is made by finding a compromise management according to the principles [28]. The components of the objective function and the constraints of the problem (3) - (8) are nonlinear and integer. This determines the choice of solution methods.

5. Operational issues by transfer traffic

Among the many aspects of port-station cooperation, the basis is the train schedule. There is a method of volumetric-calendar planning and production schedule drafting – network planning, which is related to serial or serial-parallel execution of certain works and operations in order to shorten the total delivery cycle of intermodal freight in containers [27]. One of the main instruments for organizing and technological basis for the work of all railway departments is the schedule of trains, which on the landfills of interaction with cargo-handling complexes should provide for docking:

- traffic schedule with allocation of firm schedules of freight trains on sites;
- «Christmas» schedule of the movement of transfer trains in a transport hub linking the pre-port sorting station of the junction, individual port stations and their district parks;
- contact schedule for handling vehicles.

At the same time, the «Christmas tree» schedule should provide for:

1) the correspondence of the linked threads of the core of the trains on the sections and in the junction;

2) intervals of arrival and departure of trains from the line to the junction and from the junction to the line, ensuring the possibility of following the same strings as transit for the pre-port sorting station, and processed (formed) trains with minimal delays;

3) the possibility of moving transfer locomotives between port stations (district parks, terminals) and skipping group transfers with a detachment at port stations.

There are the following types of transfer trains in the port junctions (see Figure 1):

- between the pre-port sorting station and the industrial stations (loading and unloading fronts) serving the port terminals;
- between the pre-port sorting station and the district parks;
- between the pre-port sorting station and the port stations of the junction;
- between the pre-port sorting station, port stations and district parks.

The establishment of optimal parameters for the transfer movement, namely the number of trains and the size of their trains, is based on the following recommendations. For the selected sorting distribution option in the junction, it is necessary to determine the number of wagons transferred when distributing flows from one unit of the junction to another, and for these trains to establish the optimal dimensions of the movement of the transfer trains, as well as the required number of locomotives for servicing them.

The optimal dimensions of the transfer motion are determined by the number of transfer trains, in which the total average daily costs will be minimal for the junction as a whole.

The considered order of justification of the technology of the transfer movement in the junction includes:

- determination of average daily expenses for the transfer movement for each intermodal assignment;
- determination of the required fleet of transfer locomotives;
- comprehensive selection of rational sizes of transmission motion for the junction as a whole;
- drawing up the junction train schedule.

Possible options for the size of the movement of transfer trains for each destination are limited to the minimum number of trains:

$$\mathbf{n}^{\min}{}_{\mathrm{pow}} = \frac{\mathbf{N}_{\mathrm{pow}}}{\mathbf{m}_{\mathrm{tra.max}}} \tag{9}$$

where $\mathbf{N}_{\mathrm{pow}}$ is the power of the assignment of the transfer trains, taking into account the wagon flows, distributed with the PSS at the PS or DP, wagons / day;

IOP Conf. Series: Materials Science and Engineering **795** (2020) 012005 doi:10.1088/1757-899X/795/1/012005

$\mathbf{m}_{\mathbf{tra.max}}$ – the maximum composition of the transfer train, wagons.

Comparable in different variants, the average daily cost for the transfer movement in the port railway junction is determined for each destination of the transfer trains according to the formula:

$$\mathbf{E}_{\mathbf{tran}} = \mathbf{E}_{\mathbf{accu}} + \mathbf{E}_{\mathbf{form}} + \mathbf{E}_{\mathbf{wait.dep}} + \mathbf{E}_{\mathbf{move}} + \mathbf{E}_{\mathbf{wait.arr}} + \mathbf{E}_{\mathbf{disb}} + \mathbf{E}^{\mathbf{maint}}_{\mathbf{loco}} + \mathbf{E}_{\mathbf{back}} \quad (10)$$

where $\mathbf{E}_{\mathbf{accu}}$ – the costs associated with the accumulation of transfer trains, rubles / day;
$\mathbf{E}_{\mathbf{form}}$ – expenses for the formation of transfer trains, rubles / day;
$\mathbf{E}_{\mathbf{wait.dep}}$ – costs associated with the expectation of operations with the transfer train at the departure station (the end of formation, processing and departure), rubles / day;
$\mathbf{E}_{\mathbf{move}}$ – expenses for movement of transfer trains on the intra-junction line, rubles / day;
$\mathbf{E}_{\mathbf{wait.arr}}$ – costs associated with the expectation of operations at the station of arrival of transfer trains (technological processing upon arrival and disbandment), rubles / day;
$\mathbf{E}_{\mathbf{disb}}$ – expenses for the disbandment of transfer trains, rubles / day;
$\mathbf{E}^{\mathbf{maint}}_{\mathbf{loco}}$ – expenses for the maintenance of the fleet of transfer locomotives, rubles / day;
$\mathbf{E}_{\mathbf{back}}$ – costs for reserve run of transfer locomotives due to traffic disparity, rubles / day.

To determine the optimal dimensions of the motion, you must determine each of the cost elements that is included in (10), and select the value of the $\mathbf{E}_{\mathbf{gress}}$ corresponding to the minimum value.

With the increase in the number of transfer trains, the loading of the shunting locomotives of the formation increases, the loading of the brigades of the technical service of cars at departure, the loading of the output section, the loading of the brigades of the technical service of cars for receiving, and the loading of the sorting device of the destination system. This generally affects the wagons' downtime of transfer trains at gantry stations. Therefore, these issues should be resolved in conjunction with the issues of technology and the structure of the dispatching management in the port railway junction with the concentration of work management in the hands of one replacement manager.

The behavior of the above cost elements (10) can vary considerably depending on the weight and length of transfer trains (Figure 3).

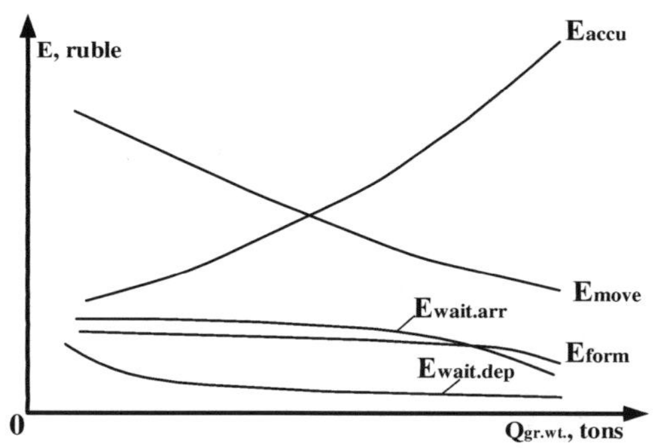

Figure 3. Dependence of the components of function (10) on the mass of the transfer train

The increase in train accumulation costs is due to increased downtime of wagons on sorting tracks, while traffic costs are reduced as a result of costs reduction associated with the mileage of the rolling stock, as well as savings on fuel or electricity costs.

6. Conclusions

1. Developed and scientifically substantiated the theoretical basis of the method of distribution of sorting and transfer work in the interaction of stations in the pre-port railway hubs.

2. The proposed classification of constraints and controlled variables in solving the problem takes into account the fluctuations of the jets and the unevenness of incoming traffic flows to the port area and the technically permissible load level of railway infrastructures.

3. The search for a solution to the problem involves directional sweeping of the wagon traffic flows into the junction, checking for constraints in accordance with the device interaction scheme (Figure 1)

and then finding the optimal variant of the distribution of sorting and transfer work among the stations at the pre-port junction.

4. The software created on the basis of the methodology will make it possible to calculate options for servicing seaports by railway transport, both with the existing capacity of permanent devices of railway facilities, and when the technical equipment of the port infrastructure is changed with the least amount of time.

5. The numerical implementation of the developed methodology made it possible to determine the best option for servicing the sea terminals of the Aktau port by rail. The evaluation of service options by the formulas (3) and (10). It was found that in the implementation of the optimal variant, a simple carriage to the port at the railway stations Aktau-port, Mangyshlak, Stroitelnaya, in the «Khimichesky» and «Port» shunting parks - decreased by an average of 8%.

6. The presence of such a technique allows: to establish a rational distribution of rolling stock, shunting means, rational implementation of sorting and transfer work; to adjust the plan for the formation of trains in the area of the pre-port railway junction, both on-line and during planning; establish performance indicators of railway stations and district (shunting) parks at junctions; evaluate the performance of the rolling stock and technical means of railway devices.

7. On the basis of the research results reviewed in this article, changes and additions have been made to the Instructive Instructions [7] regarding the calculation methodology for the distribution of sorting work and the organization of transfer traffic in the pre-port railway hubs.

References

[1] Notteboom T and Rodrigue J P 2009 *Transp. and Commun. Bulletin for Asia and the Pacific.* **78**
[2] Wilmsmeier G, Monios J and Lambert B 2011 *Journal of Transp. Geography.* **19** **6** 1379-86
[3] Krasnoshchek A, Borodin A and Rybin P 2014 *Railway transp.* **11** 26-34
[4] Krasnoshchek A, Kuznetsov A, Kuzin V, Panin V, Suslova M and Zalutskiy M 2015 *Railway transp.* **9** 8-15
[5] The customs code of Aktau international sea trade port 2016 **646**
[6] Apatsev V, Borodin A and Borodina E 2003 *Transportation management in railway junctions* (Moscow: RGOTUPS)
[7] Instructive instructions on the organization of railroad car flows on the railways of JSCo «Russian Railways» 2007 (Moscow: Tehinform).
[8] Nicholson A, Schmöcker J, Bell M and Iida Y 2003 *The network Reliability of Transp.,, Proc. of the 1st Inter. Symp. of Transp. Network Reliability*
[9] *Rail competition. Special eurobarometer 388. Report.* 2012
[10] Novikov P 2008 PhD thesis. (Ekaterinburg: UrGUPS)
[11] Baublys A 2002 *Comp. Model. & New Techn.* **6** **1** 72-81
[12] Kostenko N and Zubkov V 2007 *Far Eastern roads and Transp. Corridors* **4** 10-13
[13] Ahmedov R 2013 *Modern Scien. Research and Innov* **8**
[14] Komovkina N 2013 PhD thesis (Sankt-Petersburg: PGUPS)
[15] Kamyshova Y and Zubkov V 2015 *The Engineer's Herald of the Don* **2**
[16] Chislov O, Bogachev V, Zaporozhsky V and Bogachev T 2016 *News of PGUPS* **3** 302-313
[17] Larsen A 2008 Master Thesis (Dep. of Transp. Tech. Univ. of Denmark Kongens Lyngby)
[18] Skalov K, Zubkov M, Kravchenko V, Nikitina V and Persianov V 1965 *Port junctions and stations (Device and operation)* (Moscow: Transport)
[19] Rules and technical standards for the design of stations and junctions on railways of gauge 1520 mm 2001 (Moscow: Tehinform)
[20] Hristovskaya E 1993 PhD thesis (Moscow: MIIT)
[21] Osipov A and Khristovskaya E 1992 *Vestnik VNIIZHT* **3** 8-11
[22] Osipov A and Khristovskaya E 1992 *Railway Transp.* **5** 8-12
[23] Borodina E and Bilenko G 1995 *Transp. Science, techn., manag.* **12** 17-21
[24] Milinkovic S, Markovic S and Veskovic M 2013. *Simul. Model. Practice and Theory* **33** 144-157

[25] Li H, Jinc M and He S 2015 *Transp. Research Record: Journal of the Transp. Research Board.* **1 2475** 72–80
[26] Robinson R 2002 *Maritime Policy and Manag*. **29** 241-255
[27] Nagel K 2016 Thesis (Berlin)
[28] Borodin A and Panin V 2018 *Transp. Problems* **13 4** 37-46

The wear resistance additive products derived from EED cobalt powder

E V Ageev, A Yu Altukhov and E V Ageeva
FGBOU VO Southwest state University, 94 50 years of October street, Kursk, 305040, Russia.

E-mail: ageev_ev@mail.ru

Abstract. The main claim for powders for additive 3D technologies is a spherical shape of particles. Such particles are the most compact, packed in a certain volume and provide "fluidity" of the powder composition in the material supply systems with the least resistance. The wide use of the method of metal waste recycling into powders for their reuse and application in additive technologies is constrained by the lack of scientific and technical literature complete information on the impact of the original composition, modes and environment on the powder properties and practical application technologies. Therefore, comprehensive theoretical and practical researches are required to develop technologies for the reuse of electro-erosion powders and to estimate the effectiveness of their use. The aim of the work was to study the wear resistance of additive product samples, obtained from cobalt-chromium electro-erosion powder. The wear resistance of the samples was studied on a high-temperature tribometer manufactured by CCM tools. The friction coefficient (initial, maximum and average) is determined on the basis of the wear resistance assessment. The wear coefficient value of the statistical partner and the wear coefficient of the electric spark coating were also identified. The visual image of the wear trace and the wear groove profile of the electro-spark coating are shown.

1. Introduction

The technology of "three-dimensional printing" appeared in the late 80-ies of the last century. The pioneer in this field is the company 3D Systems, which designed the first commercial stereolithographic machine - SLA - Stereolithography Apparatus (1986).

The widespread use of digital technologies in the field of CAD, modeling and calculation (CAE) and machining (CAM) has stimulated the explosive nature of the development of printing technologies, and now it is extremely difficult to indicate the area of material production where to some extent printers would not have been used [1-8].

The main requirement for powders for additive 3d technologies is the spherical shape of the particles. Such particles are most compactly packed into a certain volume and ensure the "fluidity" of the powder composition in the supply systems of the material with minimal resistance. In addition, the powder should contain a minimum amount of dissolved gas. The microstructure of the powder must be uniform and finely dispersed (with a uniform distribution of phase constituents) [9, 10].

The main advantage of the technology of electroerosive dispersion (EED) is the use of waste as starting materials, which is much cheaper than the pure components used in traditional technologies. In addition, this technology allows you to obtain powders-alloys. [11, 12].

The wide use of the EED method for metal waste recycling into powders for the purpose of their reuse and application in additive technologies is helded back in the scientific and technical literature of full-fledged information on the effect of the initial composition, regimes and media on the properties of powders and technologies of practical application. Therefore, in order to develop technologies for the reuse of electroerosive powders and to evaluate the effectiveness of their use, complex theoretical and experimental studies are needed.

The average particle size of powder obtained by EED depends upon the pulse (arc) energy, which in its turn, depends upon the voltage across the reactor electrodes (EED plant supply voltage), holding capacity of the energy discharge capacitors, sparking voltage of the working fluid, geometric parameters of the reactor (the distance between electrodes), dimensions of the dispersed material and its resistance to erosion.

To obtain powders of the required size, it is recommended to change the capacitance of the plant's capacitors or the voltage on the electrodes. Increasing these energy parameters increases the pulse energy and the average size of the powder particles.

The aim of the work was to conduct a phase analysis of additive products from electroerosive cobalt-chrome powders.

2. Materials and methods

To carry out the planned researches, wastes of the cobalt-chromium alloy of the brand KHMS "CELLIT" were chosen. Butyl alcohol (butanol-1) was used as a working fluid. A unit for EED of conductive materials was used for the production of cobalt-chromium powders,. Dispersion parameters: voltage 100 V, capacity 48 μF, repetition rate 120 Hz.

The friction coefficient and wear rate of the surface of the samples and counterbody were measured on an automated friction machine (Tribometer, CSM Instruments, Switzerland) controlled by a computer (Figure 1), according to the standard test procedure "ball-disk" (Figure 2). These tests allow using of the Hertz model; they correspond to international standards ASTM G99-959 DIN50324 and can be used to evaluate the wear resistance of the sample and the opposing member.

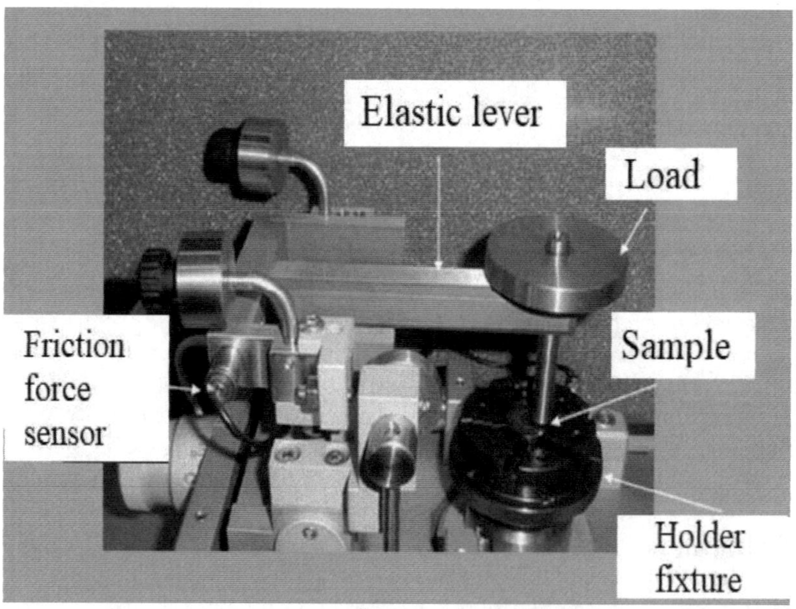

Figure 1. General view of an automated friction machine (Tribometer, CSM Instruments, Switzerland).

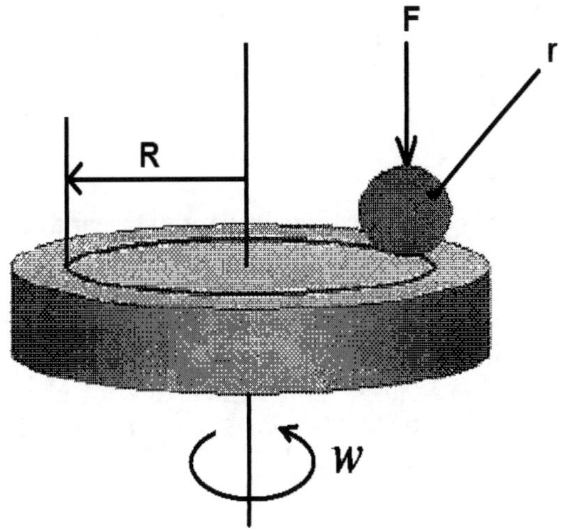

Figure 2. Standard test scheme "ball-disk", R is the radius of curvature of wear; r is the radius.

The samples were placed in a holder, a rod was attached perpendicular to the plane of the sample, at the end of which there was a ball with a diameter of 6 mm Al_2O_3 (aluminum oxide). By adjusting the displacement sensor, the radius of wear curvature was selected; another sensor compensated for the frictional force and allowed to know the value of the friction coefficient at a certain time.

Preparation for the test included:

A) Three types of calibration:

1) calibration motor speed and rotation; 2) calibration of the tangential displacement of the sensor; 3) the radius calibration.

B) test parameters setting with special software (Instrum X for Tribometer program). The following information required for the test was specified:

1) the frequency of the sensor interrogation; 2) data on the environment; - temperature, - humidity; 3) the load value, at which the test will be carried out, 4) the linear velocity, cm / sec; 5) mileage in meters or number of cycles; 6) information about the substrate: - coating material; - substrate material; - type of preliminary cleaning of the sample before testing; 7) information on the counter (ball): - coating material; - material of the counterbody; - type of preliminary cleaning of the counterbody before the test; - the size, mm; - geometry.

The tests were carried out in air at a load of 5 N and a linear velocity of 10 cm/sec, a curvature of the wear was 5-6 mm, friction path length was 1000 m.

According to test results, the wear resistance of the sample and the static partner (ball) was estimated from the wear factor by the formula:

$$W = V/(P \cdot l), \qquad (1)$$

where W is the wear rate, $mm^3 \cdot N^{-1} \cdot m^{-1}$, V is volume of deleted material, mm^3, P is load, N; l is friction path, m.

Having determined wear diameter of the ball with the Olympus GX 51 optical inverted microscope, the volume of the removed material on the ball was calculated by the following formula:

$$V = \pi \cdot h^2 \cdot (r - (1/3)h), \qquad (2)$$

where $h = r - \left(r^2 - [d/2]^2\right)^{1/2}$, d is wear diameter, mm; r is ball radius, mm; h is segment height, mm.

IOP Conf. Series: Materials Science and Engineering **795** (2020) 012006 doi:10.1088/1757-899X/795/1/012006

The volume of the removed material was determined from the section of the wear track on the surface of the sample using an automated precision contact profilometer Surtronic 25 manufactured by Taylor Hobson. The volume of the removed sample material was determined by the formula:

$$V = s \cdot l \quad (3)$$

where l is circumference, mm; s is cross-sectional area of the wear track, mm^2.

3. Results and Discussion

Visual image of the counterbody (ball) wear after multiple passes through the experimental surface of the experimental sample is shown in Figure 3.

Electron microscopic image of an additive sample after several ball passes is shown in Figure 4.

The images were obtained with the OLYMPUS GX51 inverted optical microscope, counter body was made from Al_2O_3 (aluminum oxide).

The groove profile of the surface was investigated on an automated precision contact profilometer SURTRONIC 25 (Figure 5).

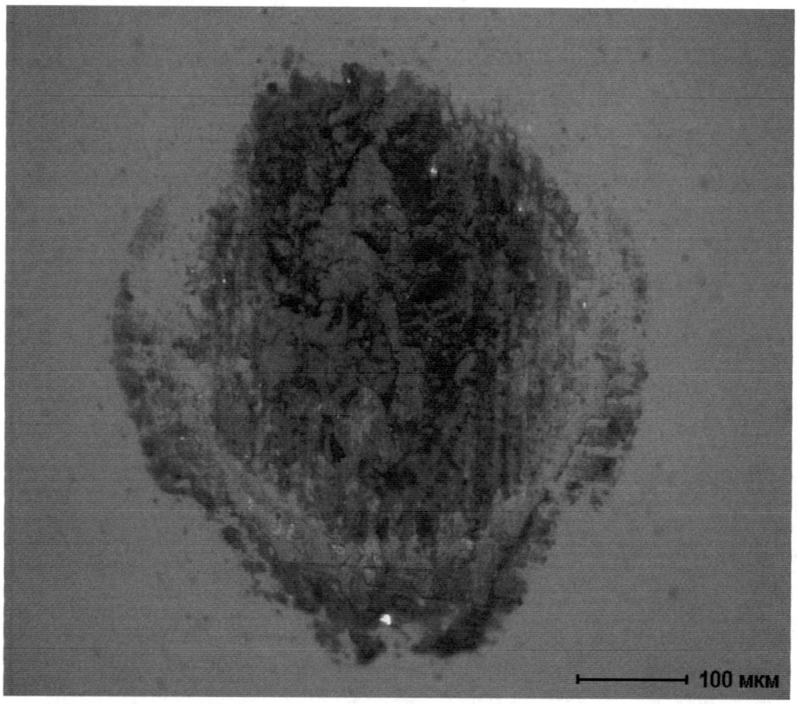

Figure 3. Image of the counterbody (ball) wear after multiple passes through the investigated sample surface.

Figure 4. Electron microscopic image of an additive sample after several ball passes.

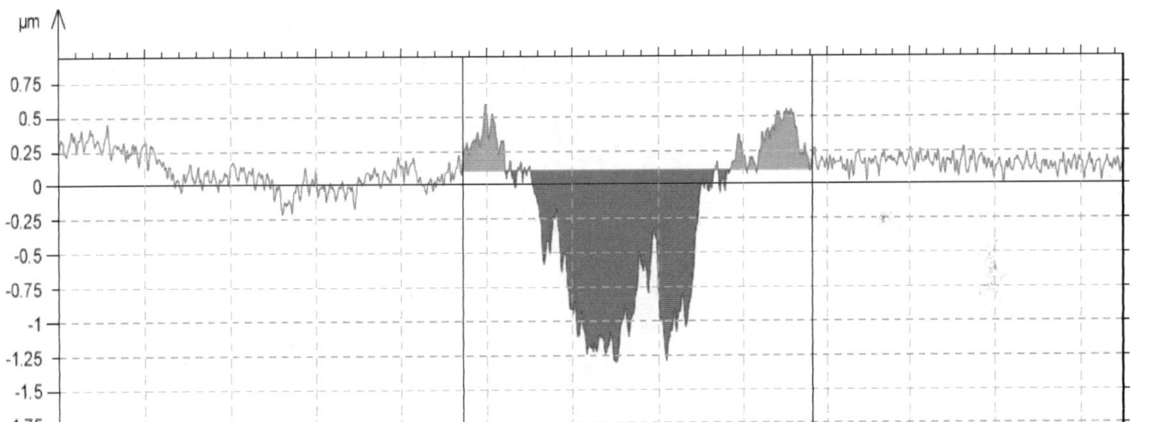

Figure 5. Wear depth on sample surface.

Based on the results of experimental studies, the following characteristics of wear resistance were obtained: the average value of the friction coefficient 0.63; the wear factor of the static partner 0.104; the wear factor of the sample 3.37.

Based on the evaluation of the wear resistance, the friction coefficient (initial, maximum and average), the values of the wear factor of the static partner and the wear factor of the electro-spark coating were also obtained. The optical image of the wear trace and the groove profile of the wear of the electro-spark coating are shown.

4. Conclusion

According to the results of the studies of the wear resistance of additive products from electroerosive cobalt-chromium powders, it was experimentally established that the average value of the friction coefficient is about 0.673.

The conducted research allows using progressive, environmentally friendly, low-tonnage and waste-free technology of electroerosive dispersion to obtain new powder materials from waste of cobalt-chrome alloys and additive products from them with guaranteed wear resistance.

Acknowledgments
The work was supported by the Russian Science Foundation. Project number is 17-79-20336.

References

[1] Wang Z, Guana K, Gaoa M. 2012 *Journal of Alloys and Compounds* **513** 518-523.
[2] Biamino S, Penna A, Ackelid U 2011 *Intermetallics*. **19** 776-781.
[3] Gu D D, Meiners W, Wissenbach K, Poprawe R. 2012 *International Materials Reviews* **57** 133-164.
[4] Song B, Dong S, Zhang B 2012 *Materials & Design* **35** 120-125.
[5] Song B, Dong S, Coddet P 2012 *Surface and Coatings Technology* **206** 4704-4709.
[6] Wang Z, Guana K, Gaoa M 2012 *Journal of Alloys and Compounds* **513** 518–523.
[7] Karlsson J, Snis A, Engqvist H, Lausmaa J 2013 *Journal of Materials Processing Technology* **213** 2109-2118.
[8] Safdar A, He H Z, Wei L Y, Snis A 2012 *Rapid Prototyping Journal* **18** 401–408.
[9] Loeber L, Biamino S, Ackelid U 2011 *Solid freeform fabrication proceedings* **11** 547-556.
[10] Gu D D, Meiners W, Wissenbach K, Poprawe R. 2012 *International Materials Reviews* **57** 133-164.
[11] Ageeva E V, Ageev E V, Karpenko V Yu, Osminina A S 2014 *Journal of nano- and electronic physics* **3** 03049-1– 03049-3.
[12] Ageeva E V, Ageev E V, Horyakova N M, Malukhov V S 2014 *Journal of nano- and electronic physics* **3** 03011-1 - 03011-3.

IOP Conf. Series: Materials Science and Engineering **795** (2020) 012007 doi:10.1088/1757-899X/795/1/012007

Obtaining and Research of Additive Products from Electro-erosive Cobalt-Chrome Powders

E V Ageev, A Yu Altukhov and E V Ageeva
FGBOU VO Southwest state University, 94 50 years of October street, Kursk, 305040, Russia.

E-mail: ageev_ev@mail.ru

Abstract. Based on the particularity of the methods of obtaining spherical shape powders in order to obtain spherical granules of a regulated grain size, the technology of electroerosive dispersion, which is distinguished by relatively low energy costs and ecological cleanliness of the process, is described. The widespread use of the method of EED for the processing of metal waste into powders for the purpose of their reuse and application in additive technologies is constrained by the lack of complete information in the scientific and technical literature about the influence of the original composition, modes and media on the properties of powders and technologies of practical application. Therefore, the development of technologies for the reuse of EED powders and the evaluation of the effectiveness of their use requires the conduct of comprehensive theoretical and experimental studies. The purpose of this work was to obtain and study additive products from electroerosive cobalt-chromium powders of a specific particle size distribution and to study their microstructure. The granulometry of the obtained powders was determined on a laser analyzer of particle sizes "Analysette 22 NanoTec". The microstructure of additive samples from cobalt-chromium powders was investigated by optical microscopy on an inverted optical microscope OLYMPUS GX51. On the basis of completed studies, aimed at obtaining and studying additive products from electroerosive cobalt-chrome powders of a specific particle size distribution, and studying their microstructure, it was found that additive samples, obtained from a cobalt-chrome powder with an average particle size of 35.68 μm, have practically no pores.

1. Introduction

At present, manufacturers of additive machines are also suppliers of expendable materials, and these firms themselves do not produce powders, but participate in their preparation for work on their machines. Powder is purchased from manufacturers of powder materials that produce material for the diverse needs of powder metallurgy. The purchased powder is subjected to screening and separation into fractions, the fractions are mixed in certain ratios and packed in an airtight container. Thus, in particular, manufacturers of additive machines insure themselves against possible claims to the quality of powders on the part of consumers. Abroad, issues of standardization of materials for additive technologies are the responsibility of NIST - National Institute of Standards and Technology (USA); International Organization for Standardization ISO (Committee TC261 on additive technologies), and ASTM (International Committee on Additive Technologies) [1-8].

For domestic consumers, expendable materials are a serious problem. Due to the underdevelopment of the Russian market, metal powder compositions for additive machines generally have to be purchased abroad.

The main request for powders for additive machines is the spherical shape of the particles. Such particles most compactly fit into a certain volume and provide the "fluidity" of the powder composition in the material supply systems with minimal resistance. In addition, the powder must contain a minimum amount of dissolved gas. The microstructure of the powder should be homogeneous and fine (with a uniform distribution of phase components) [9, 10].

Based on the particularities of the methods of obtaining spherical powders in order to obtain spherical granules of a regulated grain size, the technology of electroerosive dispersion, which is distinguished by relatively low energy costs and ecological cleanliness of the process, is proposed [11, 12].

The main excellence of the proposed technology is the use of waste as primary products, which is much cheaper than the pure components used in traditional technologies. In addition, this technology is powder, which allows to obtain powder-alloys.

The widespread use of the method of EED for the processing of metal waste into powders for the purpose of their reuse and application in additive technologies is constrained by the lack of complete information in the scientific and technical literature on the influence of the original composition, modes and media on the properties of powders and technologies of practical application. Therefore, the development of technologies for the reuse of EED powders and the evaluation of the efficiency of their use requires the conduct of comprehensive theoretical and experimental studies.

The purpose of this work was to obtain and study additive products, fabricated from electroerosive cobalt-chromium powders of a specific particle size distribution and to study their microstructure.

2. Materials and methods

To carry out the planned research, the waste of a cobalt-chromium alloy of the grade KHMS "ZELLIT" was selected. Butyl alcohol was used as a working fluid. To obtain cobalt-chromium powder materials by electro-erosion dispersion, an installation for EED of conductive materials was used. The dispersible material is poured into a desiccator, which is filled with butyl alcohol, which is used as a working fluid for dispersion.C_4H_9OH butyl alcohol is representative of monohydric alcohols. It is a colorless, slightly viscous liquid, with no color and a characteristic smell of fusel oil. It is mixed with organic solvents. Butanol is used as a solvent in the paint industry, in the manufacture of resins and plasticizers, to produce plasticizers, as well as in many other industries.

To obtain experimental samples, an installation for layer-by-layer deposition of powder materials by plasma was used (Figure 1).

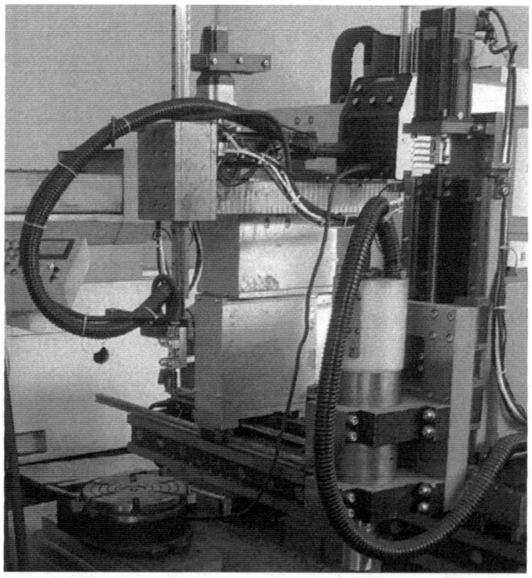

Figure 1. Plant for powder coating.

IOP Conf. Series: Materials Science and Engineering **795** (2020) 012007 doi:10.1088/1757-899X/795/1/012007

The granulometry of the obtained powders was determined on a laser analyzer of particle sizes "Analysette 22 NanoTec". This device defines the size allocation of particles in suspensions, emulsions and aerosols. Compared to "classical" measurement methods — sieving, sedimentation or image analysis — laser diffraction has several significant benefits, such as short analysis time, good reproducibility and exactness, simple calibration, large measurement range and high versatility. The measurement range of the "Analysette 22 NanoTec" is from 0.01 to 2000 μm. The analyzers, which determine the particle size distribution by laser diffraction, use the physical principle of the scattering of electromagnetic waves. The design consists of a laser, through a measuring cell directed to the detector. With the help of a dispersing device, the particles are fed into the measuring cell and pass through the laser beam. Light scattered in proportion to the particle size is focused onto the detector through a lens. Using the distribution of scattered light using complex mathematics, the distribution of particles by their sizes is calculated. The result is a volume fraction appropriating to equivalent diameters in laser diffraction. Thanks to the built-in ultrasonic bath (volume about 500 ml, energy and ultrasound frequency 80 W / 36 kHz); even hard-to-disperse samples can be analyzed without the use of extra equipment. The digital ultrasonic generator always keeps the installed power at an optimum and constant level. The lower limit of sensitivity with small quantities of small and large particles in their size distributions (within the measurement range) is 3%. Reproducibility according to ISO 13320-1 $d_{50} \leq 1\%$.

The microstructure of additive samples from cobalt-chromium powders (by transverse polishing) was investigated by optical microscopy on an inverted optical microscope OLYMPUS GX51. The surface (by grinding) of the samples was ground and polished. Grinding produced metallographic paper with a large (№№ 60-70) and fine grain (№ № 220-240). During the grinding process, the sample was serially rotated 90 °. The abrasive particles were washed with water and subjected to polishing on the wheel with suspensions of metal oxides (Fe_3O_4, Cr_2O_3, Al_2O_3). After reaching the mirror gloss, the area of the thin section was washed with water, alcohol, and dried with filter paper.

The inverted optical microscope OLYMPUS GX51 - research microscope is designed to receive in reflected light: bright field and dark field images; differential interference contrast (DIC) images; images in polarized light. The maximum magnification of the microscope: × 1000 (interchangeable lenses × 5, × 10, × 20, × 50, × 100). Besides, the OLYMPUS GX51 microscope is equipped with the PS 11 precision scanning automated table, the SIMAGIS 2P-2C digital microscope video camera and the SIAMS Photolab automated image analysis system (SIMAGIS Research) and ImageScope M. The system of image analysis "SIAMS Photolab" is a software product, designed for processing and analyzing images, obtained using digital and analog photo and video cameras, as well as scanners during micro and macro photography. Image processing in "SIAMS Photolab" is executed in a chain of interconnected cells containing the original image, the results of intermediate processing steps, the final processed image and the measurement results in the form of numbers, graphs and histograms.After a chain formation for processing a new sample according to a given algorithm, it is enough to replace the original image. At the same time, the user can visually control and manually regulate the parameters of any processing stage. In addition to automated processing, the system allows to edit images in manual and semi-automatic methods. The system provides for generating reports in MS Word format and exporting images, numeric and text data to the most common formats.

3. Results and Discussion

The granulometric composition of the cobalt-chrome powder, obtained by the electro-erosive dispersion of the waste material of the KHMS alloy, is presented in Figure 2.

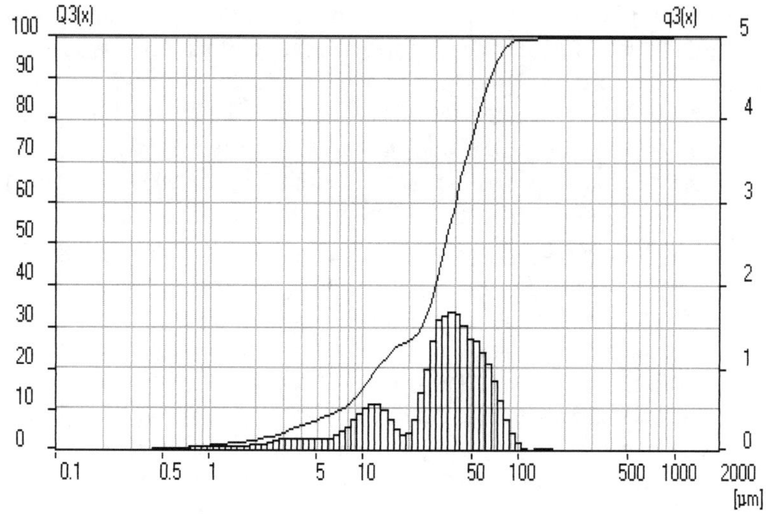

Figure 2. Size distribution of powder microparticles.

In Figure 2, the integral curve and histogram: integral curve in coordinates Q3 (x) = f (µm) (left scale) - each point on the curve indicates how much percentage of the sample has a particle size less than or equal to this. The histogram in coordinates q3 (x) = f (µm) (right scale) - the number of sample with a given particle size.

It has been experimentally established that the average particle size of an EDM cobalt-chrome powder is 35.68 microns, the arithmetic value is 35.679 µm.

The microstructure of additive samples from cobalt-chromium powders (by transverse polishing), studied by optical microscopy on an inverted optical microscope OLYMPUS GX51, is shown in Figure 3.

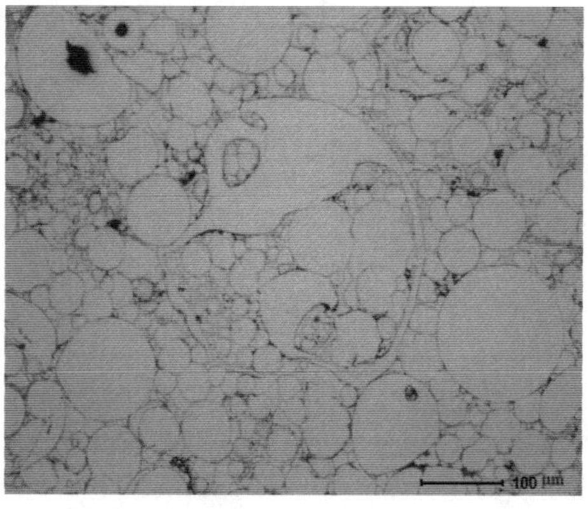

Figure 3. Microstructure of the additive sample.

It was established experimentally that the microstructure of additive samples of particles of electroerosive cobalt-chrome powder has practically no pores.

4. Conclusion

The main claim for powders for additive machines is the spherical shape of the particles. Such particles most compactly fit into a certain volume and provide the "fluidity" of the powder composition in the material supply systems with minimal resistance. Based on the particularity of the methods of obtaining spherical shape powders in order to obtain spherical granules of a regulated grain size, the technology of electroerosive dispersion, which is distinguished by relatively low energy costs and ecological cleanliness of the process, is described. The main advantage of the proposed technology is the use of waste as primary materials, which is much cheaper than the pure components used in traditional technologies. In addition, this technology is powder, which allows to obtain powder-alloys. According to the results of the conducted studies it was experimentally established that the microstructure of additive products obtained from electroerosive powders with an average particle size of 35.68 microns has practically no pores.

Acknowledgments
The work was supported by the Russian Science Foundation. Project number is 17-79-20336.

References
[1] Karlsson J, Snis A, Engqvist H, Lausmaa J 2013 *Journal of Materials Processing Technology* **213** 2109-18.
[2] Safdar A, He H Z, Wei L Y, Snis A 2012 *Rapid Prototyping Journal* **18** 401–408.
[3] Loeber L, Biamino S, Ackelid U 2011 *Solid freeform fabrication proceedings* **11** 547-556.
[4] Gu D D, Meiners W, Wissenbach K, Poprawe R. 2012 *International Materials Reviews* **57** 133-164.
[5] Wang Z, Guana K, Gaoa M. 2012 *Journal of Alloys and Compounds* **513** 518-523.
[6] Biamino S, Penna A, AckelidU 2011 *Intermetallics*. **19** 776-781.
[7] Gu D D, Meiners W, Wissenbach K, Poprawe R. 2012 *International Materials Reviews***57** 133-164.
[8] Song B, Dong S, Zhang B 2012 *Materials & Design* **35** 120-125.
[9] Song B, Dong S, Coddet P 2012 *Surface and Coatings Technology* **206** 4704-09.
[10] Wang Z, Guana K, Gaoa M 2012 *Journal of Alloys and Compounds* **513** 518–523.
[11] Ageeva E V, Altukhov A Yu, Khardikov S V, Gulidin S S, Novikov A N 2015 *Journal of nano- and electronic physics* **4** 04080(3).
[12] Ageeva E V, Ageev E V, Pikalov S V, Vorobiev E A, Novikov A N 2015 *Journal of nano- and electronic physics* **3**04058(2)

Size dependence of the dislocation segment vibration spectrum for various types of dislocations

I L Bataronov and V V Dezhin
Voronezh State Technical University, Moskovsky prospect 14, Voronezh 394026, Russian Federation

E-mail: viktor.dezhin@mail.ru

Abstract. The vibration spectra of the edge, mixed and screw dislocations, depending on the dislocation segment length, are considered. The value of Poisson's ratio 1/3 is assumed. The normalized values of the eigenfrequencies and damping coefficients for different frequency numbers by equating the inverse matrix of dislocation oscillators generalized susceptibilities to zero were found. It is seen from the dependences obtained that in the case of a screw dislocation, the transformation of the dislocation oscillators generalized susceptibility poles occurs. For edge and mixed dislocations the transformation of the poles was not observed. This agrees with the results of the vibration spectrum orientation dependence study. The authors suggest that these differences are due to the presence of a screw dislocation quasilocal vibration branches.

1. Introduction

During operation, engineering products experience various mechanical loads, including cyclic ones. Cyclic loads lead to vibrations affecting the material of mechanisms and machines. Therefore, understanding the processes occurring in the crystalline structure of the material is of great importance. As is known, dislocations (linear defects) play a decisive role in the mechanical behavior of crystals. In real crystals, dislocations are fixed on other defects; therefore, it is necessary to consider dislocation segments and their vibrations arising from vibration of structures. To describe these vibrations of dislocation segments, a string model was proposed [1], which was further developed in works [2, 3]. The string model is also currently used to describe particular questions of the dislocation segment vibrations [4-7]. It was noted in work [8] that the imperfection of the string model is the nonlocality of the dislocation characteristics. This was taken into account in papers [9–11], in which dislocation vibrations and phonon drag in crystals with a simple lattice were studied on the basis of the self-consistent dynamic theory. A generalized susceptibility [12], which is the linear response function of the dislocation to the applied force, was found in [13] to describe the bending vibrations of an infinite dislocation. In article [13], the calculation was based on the Lagrangian of a crystal with a dislocation; the string model was not used. Based on the results of [13], a generalized susceptibility of a dislocation segment was found in [14]. In [15], the orientation dependence of the dislocation segment vibration spectrum was studied. Sharp changes in the curves of the dependences and the corresponding transformation of the generalized susceptibility poles are observed. In this paper we study the size dependence of the dislocation segment vibration spectrum for various types of dislocations.

IOP Conf. Series: Materials Science and Engineering **795** (2020) 012008 doi:10.1088/1757-899X/795/1/012008

2. Level surface of the real and the imaginary part of the inverse generalized susceptibility B

The real part of the vibration spectrum $\mathrm{Re}(\omega - i\gamma) = \omega$ is the natural frequencies of the dislocation segment vibrations, and the imaginary part of the vibration spectrum $-\mathrm{Im}(\omega - i\gamma) = \gamma$ is the damping coefficients of the dislocation segment vibrations. The normalized frequencies of the vibration spectrum are studied in this work: $\tilde{\omega}_n = \mathrm{Re}\,(\omega_n - i\gamma_n)L/c_t = \omega_n L/c_t$ and $\tilde{\gamma}_n = -\mathrm{Im}\,(\omega_n - i\gamma_n)L/c_t = \gamma_n L/c_t$, where L is the length of the dislocation segment, and c_t is the velocity of transverse sound waves. The normalized lengths $\tilde{L} = \ln(e^C L/\lambda)^2$ of the segment were used, where λ is the half-width of the dislocation, and $C \approx 0.577$ is the Euler constant. The values of $\tilde{\omega}$ and $\tilde{\gamma}$ for different numbers of frequencies were found from the solution of the equation $B(\tilde{\omega} - i\tilde{\gamma}) = 0$, where $B(z)$ is the matrix of the dislocation segment inverse generalized susceptibilities. Figures 1 and 2 show the level surfaces of the matrix $B(z)$ for a screw dislocation, and the numbers (typed in bold) indicate the pole numbers of the generalized susceptibility.

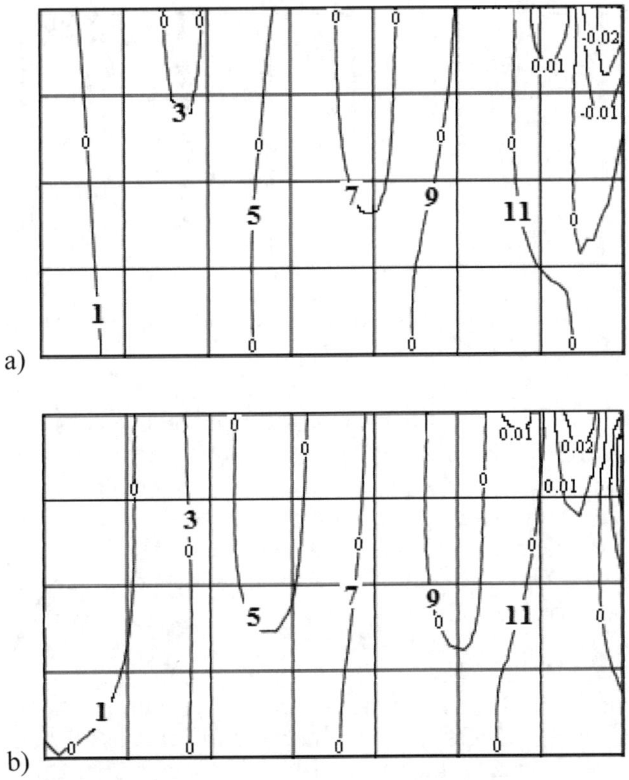

Figure 1. Level surfaces of the real (a) and imaginary part (b) of the inverse generalized susceptibility at $\tilde{L} \approx 10.4$.

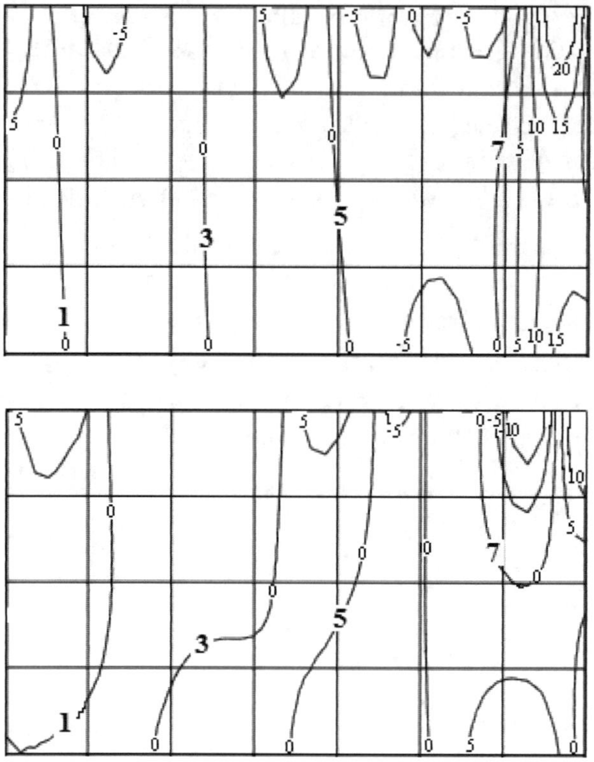

Figure 2. Level surfaces of the real (a) and imaginary part (b) of the inverse generalized susceptibility at $\tilde{L} \approx 15.0$.

For even pole numbers of the generalized susceptibility, similar graphs are obtained. The figures show that the transformation of the poles of the dislocation oscillators generalized susceptibility occurs. The pole corresponding to the third frequency goes into the region of high damping, and the pole corresponding to the fifth frequency comes in its place. For the following frequencies, a corresponding shift occurs. Then the pole corresponding to the seventh frequency leaves, and again the pole shift occurs. In total, 6 different values of the normalized length of the dislocation segment were investigated, 6 figures were obtained (figures 1 and 2 – two of them). For even frequency numbers, 6 figures were also obtained.

3. Size dependences of the dislocation segment vibration spectrum

Based on the obtained level surfaces, the dependences of the oscillation frequencies and the corresponding damping coefficients on the screw dislocation segment length are constructed (figures 3, 4). From figures 3a and 4a, it can be seen that the frequencies at $L \to \infty$ tend to the limiting values $\tilde{\omega}_n = \sqrt{2} n\pi$, since $c_d / c_t = \sqrt{2}$ for a screw dislocation [16], and the damping coefficients tend to zero (figures 3b, 4b). The dashed curve in figures 3a, 4a defines the boundary of the Debye frequencies; therefore, the points located to the left of this curve have no real physical meaning. In this paper, they are left as initial ones. Between the dashed and dash-dotted curves there is a region in which a strong dispersion of the crystal elastic vibrations should be taken into account.

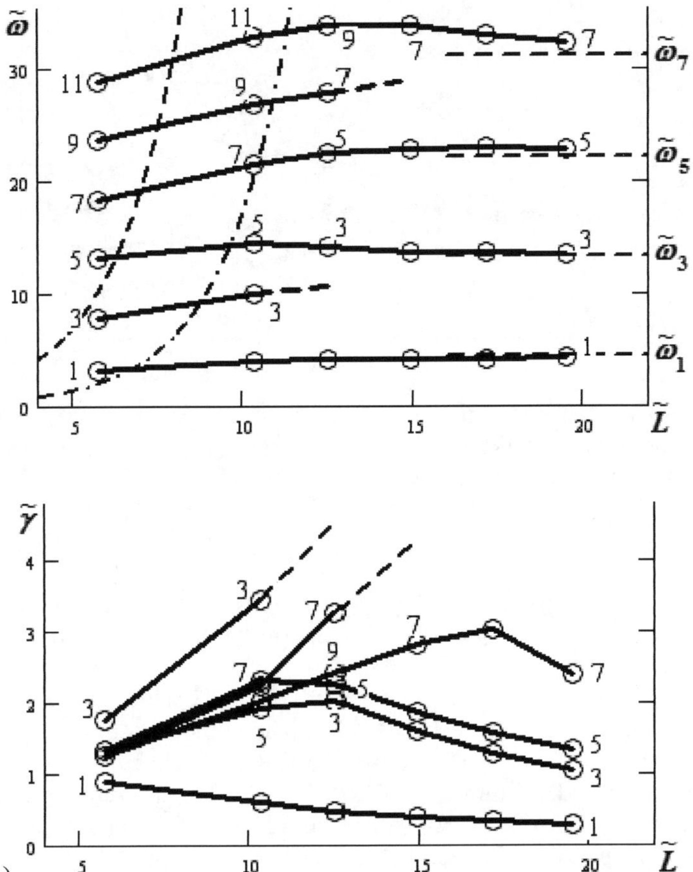

a)

b)

Figure 3. The dependence of the natural frequencies (a) and damping coefficients (b) of the screw dislocation segment vibrations on its length for odd frequencies.

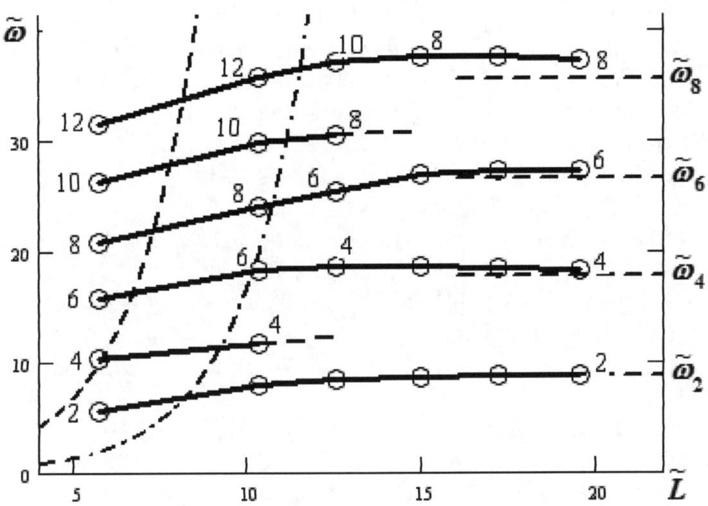

a)

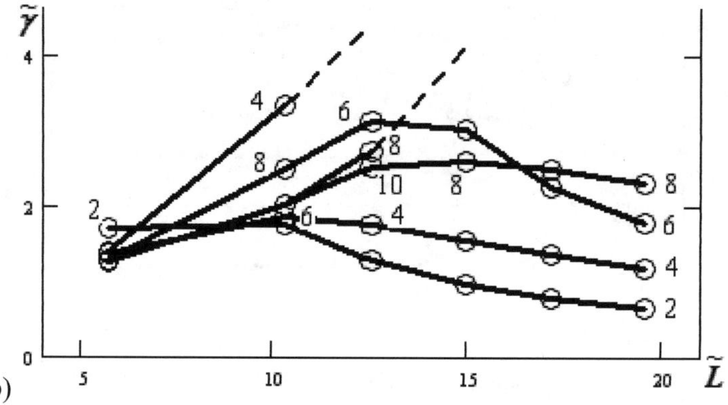

b)

Figure 4. The dependence of the natural frequencies (a) and damping coefficients (b) of the screw dislocation segment vibrations on its length for even frequencies.

4. Conclusion

In this paper, we find a size dependence of the dislocation segment vibration spectrum for various types of dislocations. The transformation of the dislocation oscillators generalized susceptibility poles, discovered in this study, was observed in the case of screw dislocation. For the edge (90°) and mixed (45°) dislocations of the pole transformation was not observed. This is consistent with the results of the vibration spectrum orientation dependence study [15]. The authors assume that these differences are associated with the presence of screw dislocation vibrations quasilocal branches [16]. The obtained results can be used for further development of the dislocation amplitude-independent internal friction theory. Also, the results can be used to study the processes occurring in the material of the mechanisms subjected to vibration influences.

References

[1] Koehler J S 1952 *Imperfections in Nearly Perfect Crystals*, ed W Schockley et al (New York: Wiley) p 197
[2] Granato A V and Lucke K 1956 *J. Appl. Phys.* **27** 583
[3] Granato A V and Lucke K 1956 *J. Appl. Phys.* **27** 789
[4] Charkina O V and Chishko K A 2001 *Phys. Solid State* **43** 1898
[5] Kamaeva O V and Chernov V M 2002 *Phys. Solid State* **44** 1676
[6] Maurel A, Pagneux V, Barra F and Lund F 2005 *Phys. Rev. B* **72** 174110
[7] Rodriguez N, Maurel A, Pagneux V, Barra F and Lund F 2009 *J. Appl. Phys.* **106** 054910
[8] Bross H and Stenzel G 1967 *Phys. Lett.* **25** 236
[9] Ninomiya T and Ishioka Sh 1967 *J. Phys. Soc. Japan* **23** 361
[10] Ninomiya T 1968 *J. Phys. Soc. Japan* **25** 830
[11] Ninomiya T 1970 *Fundamental Aspects of Dislocations Theory*, ed J A Simmons, R de Wit and R Bullough (Nat. Bur. Stand. (U.S.) Spec. Publ. 317) p 315
[12] Landau L D and Lifshitz E M 1980 *Statistical Physics* Part 1 (*Course of Theoretical Physics* vol 5) (Oxford: Butterworth-Heinemann)
[13] Bataronov I L, Dezhin V V and Roshchupkin A M 1993 *Bulletin Russian Academy Sciences - Physics* **57** 1947
[14] Bataronov I L and Dezhin V V 2016 *J. Phys.: Conf. Ser.* **738** 012108
[15] Bataronov I L and Dezhin V V 2018 *J. Phys.: Conf. Ser.* **1141** 012070
[16] Bataronov I L and Dezhin V V 2017 *J. Phys.: Conf. Ser.* **936** 012035

IOP Conf. Series: Materials Science and Engineering **795** (2020) 012009 doi:10.1088/1757-899X/795/1/012009

Relationship between phase composition and wear resistance of titanium alloys

I V Ivanov[1], A Thoemmes, K I Emurlaev, E I Tkachenko and E A Rudenya
Department of Material Science in Mechanical Engineering
Novosibirsk State Technical University, 20 Prospekt K. Marksa, Novosibirsk, Russia

E-mail: i.ivanov@corp.nstu.ru

Abstract. The wide use of titanium alloys is often due to their unique properties. Availability of a wide variety of mechanical, physical and chemical properties is associated with a large number of polymorphic modifications of titanium alloys. During the last decade, the metastable omega phase piques the largest interest among all titanium modifications. This study attempts to explain the different wear behavior of some titanium alloys with different Nb content during dry friction impact.

1. Introduction

An ideal metallic biomaterial should provide perfect biocompatibility, excellent corrosion resistance, high strength and a low Young`s modulus which is close to the Young`s modulus of human bone (4 – 30 GPa) [1-5]. The nowadays most commonly used metallic alloys for biomedical application are Cr-Co based alloys, stainless steels and titanium and its alloys [6,7]. From the beforementioned alloys titanium and its alloys are, due to their low Young's modulus the most promising candidates. Ti-6Al-4V, an alloy originally invented for aerospace application, possess a Young's modulus of 110 GPa, whereas Cr-Co alloys and stainless-steel show values minimum twice higher (240 and 210 GPa) [8, 9]. The significant mismatch between the Young's moduli of the metallic biomaterials can be the reason for the so called "stress shielding effect", which could lead to the loosening of the implant [10]. It should be mentioned that several generations of biomedical Ti alloys have been already developed and Ti-6Al-4V is in use as metallic biomaterial since decades. However, due to several reasons it is no longer considered as the perfect biomaterial. First, aluminium can be the reason for Alzheimer`s or Parkinson`s diseases [11], while vanadium is known as considerably toxic element for human body [12]. Thus, Ti alloys of new generation are currently of great interest. These alloys should possess in addition to a lower Young's modulus, better biocompatibility and an improved wear resistance. Especially the possibilities to enhance the wear resistance captured the interest of several research groups working on different Ti alloy systems (TMZF [13, 14], TNTZ [15, 16] and Ti-Nb [17]). Because the mechanical properties are mainly influenced by the alloying elements and the resulting phases a special focus should be given to the possibility to control the phase composition of the alloys. Several phases may be formed in Ti alloys. Namely equilibrium hcp α- and bcc β phase and metastable phases like hexagonal α', ω and orthorhombic α''. Depending on the processing route and the alloy composition the formation of metastable hexagonal ω phase is also possible [18, 19]. Most of

[1] To whom any correspondence should be addressed.

the abovementioned alloys consists, due to the high content of β isomorphous stabilizers, mainly of β phase, which is known to show a low Young's modulus and acceptable wear resistance.
Due to the low alloying concentration when stable β phase is observed and the absence of a miscibility gap the Ti-Nb possess a model character for new low modulus Ti-based metallic materials for biomedical application. Only a few data are available on the wear resistance of Ti-Nb alloys with different metastable phases. It can be suggested that the reason for this is the low modulus of Ti-Nb alloys consisting mainly of β phase. Therefore, this paper aims to research the influence of different phases (α, β and ω phase) on the wear resistance of some titanium alloys.

2. Experimental procedure

Biocompatible titanium alloys cp-Ti; Ti30Nb and Ti45Nb were prepared from commercially pure (c.p.) Ti and Nb by using the BUEHLER ArcMelter melting furnace. The melting processes were conducted in argon atmosphere (800 mbar) with pre-flushing. Cylindrical specimens with a diameter 10 mm and a height of 20 mm were cut from the obtained ingots. After the cutting, the samples were mechanically ground by using a 1000 grit SiC sandpaper. The final polishing was performed using a suspension consisting of 73% (vol.) colloidal SiO2 oxide, 18% (vol.) H2O2 (solution 40%) and 9% (vol.) Kroll's Solution (Kroll's Reagent). The Kroll solution consisted of 92% (vol.) H2O, 6% (vol.) HNO3, 2% (vol.) HF. After polishing cylinders have been annealed at 900°C in vacuum during 1 h.

To carry out wear experiments a friction apparatus was developed at the Materials Science department of NSTU. That allowed to varying the load and rotation speed of the material under study and found optimal conditions for wearing experiments. In order to minimize abrasion of the counterbody during the experiment, tungsten-carbide alloy (WC-Co) indenter was used. Table 1 presents some technical details of the experiment.

Table 1. Details of friction experiment. True contact area was estimated according to Profilometry images of indenter after friction experiment.

Rotation rate, rpm	**Linear friction velocity, mm/s**	**Apparent contact area, mm^2**
10	4.7	3
Real contact area, mm^2	**Indention load, kg**	**Contact pressure, MPa**
2	17.35	86.75

The surface of the alloys after friction were investigated by using ZYGO New View 7300 optical surface profilometer with 5X objective. Obtained data processing was carried out by using Python programming language.

The initial surface phase content was analyzed by synchrotron radiation. Experiments were carried out at the beamline P07 (The High Energy Materials Science) at Deutsches Elektronen-Synchrotron (DESY). The used wavelength was 0.014 nm. A 16-inch 2D PerkinElmer XRD 1621 scintillation detector with resolution of 2048 × 2048 pixels was used to record the diffraction patterns. The obtained diffraction patterns were azimuthally integrated by means of the pyFAI open-source software package [18]. The microhardness was evaluated with a WOLPERT Group 402 MVD Vickers hardness tester under a load of 0.98 N and a dwell time of 10 s.

3. Results and Discussion

It's believed that the main reason for the low wear resistance of titanium alloys is the adhesion of some sections of the material surface and the counterbody. This process is the main reason fot the appearance of debris, which upon further frictional impact can lead to active wear as abrasive. In addition, it is believed that the cause of the adhesion can be repassivation of the surface [13]. Also, in some works it is reported that the frictional resistance of titanium alloys can significantly depends on their phase composition [19].

Results of synchrotron X-ray phase analysis of the studied materials are shown in Figure 1. It can be seen that cp-Ti and Ti45Nb are single-phase materials consisting of hexagonal α- and cubic β-phases, respectively. However, Ti30Nb is characterized by a complex phase composition. This material besides α- and β-, includes hexagonal ω-phase.

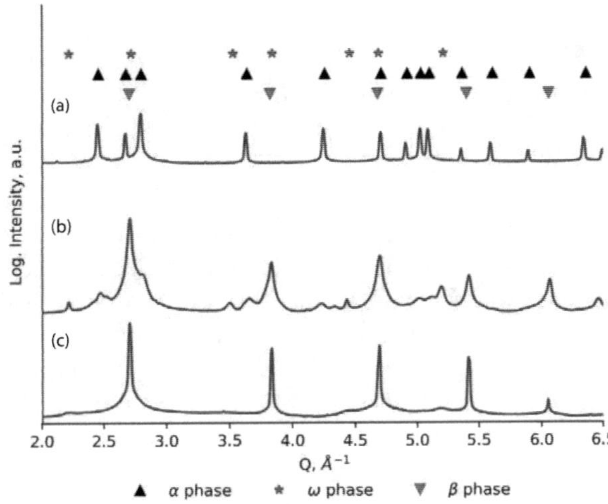

Figure 1. XRD phase analysis a – Ti; b – Ti30Nb; c – Ti45Nb.

Among the stable phases (α and β - phase) as well as among the metastable phases (α', α'' and ω - phase) in titanium alloys it is well known that the ω phases shows the most significant influence on physical and mechanical properties. Namely the ω phase increase markedly the Young's modulus, bending strength as well as the microhardness of the alloy by restricting dislocation movement [20] A high volume fraction of ω phase causes embrittlement [21]. According to the surface profile analysis after 250 cycles of friction (Figure 2), the state of the surface after impact directly depends on the phase composition of the alloy. It can be seen that the entire surface of α-titanium has undergone significant wear. Both grooves (blue) and areas of the most deformed material located above the median plane (red) of the surface are visible on the surface.

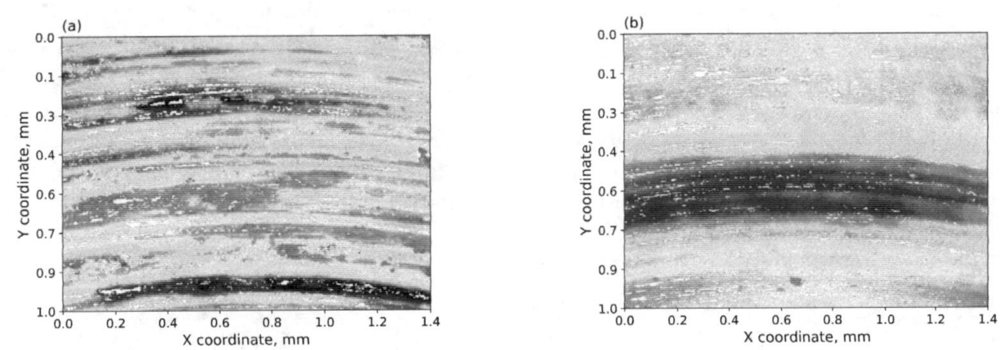

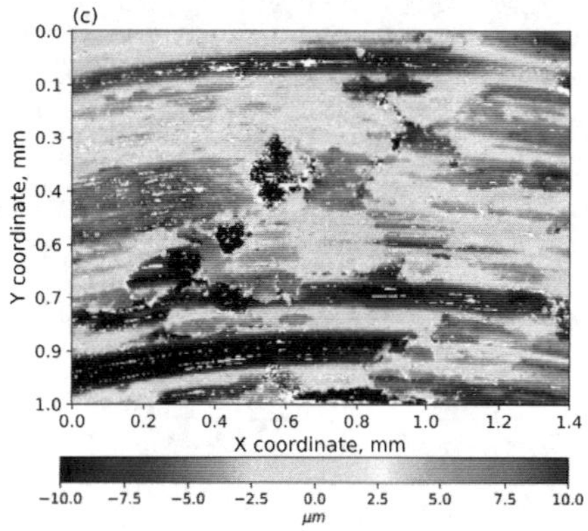

Figure 2. Surface of: a – Ti; b – Ti30Nb; c – Ti45Nb after 250 friction cycles

The results of the adhesion process — tearing off and shearing of the surface layers of the material with the formation of debris are clearly visible on the β-titanium surface. According to the surface roughness analysis (Table 2), the value of R_a, (arithmetical mean of the absolute values of profile high in a sampling surface) reaches a maximum in the Ti45Nb sample among all investigated materials. Krishnan H. et al. [19] attribute this behavior of β-titanium to the fact that, during friction the β phase is not strengthened. Detailed study of this behavior requires in-situ experiments of structure changing of the surface layers.

Table 2. Roughness parameters of titanium alloys after 250 and 500 cycles of friction.

250 cycles	**Ti**	**Ti30Nb**	**Ti45Nb**
Ra	3.08	3.36	4.5
Rz	18.33	13.46	24.32
500 cycles	**Ti**	**Ti30Nb**	**Ti45Nb**
Ra	3.64	2.27	11.52
Rz	18.14	11.75	59.72

However, in the case of Ti30Nb the delaminated surface areas were not observed. It can be seen that in the process of friction the indenter contacted the material mainly in only one region 0.3-0.6 mm wide. It is likely that this behavior is associated with increased mechanical properties of the ω -phase. As mentioned before, this phase is the hardest and strongest among all the phases of titanium. Table 3 shows the microhardness of the investigated alloys. It can be seen that the presence of the ω -phase significantly increases the microhardness of the material. Thus, the presence of one wide groove can be explained by the increased microhardness of Ti30Nb compared to Ti and Ti45Nb. It is likely that during the initial stages of the friction, the counterbody contacted only a very small area of the material and due to the high hardness did not penetrate significantly deeper into the material. As a result, the wear of Ti30Nb occurred much slower than Ti and Ti45Nb. This assumption is confirmed by the study the surface after 500 friction cycles (Figure 3).

Table 3. Vickers microhardness of titanium alloys.

	Ti	**Ti30Nb**	**Ti45Nb**
$HV_{0.1}$	132±15	298±10	231±14

IOP Conf. Series: Materials Science and Engineering **795** (2020) 012009 doi:10.1088/1757-899X/795/1/012009

After 500 friction cycles, the surface of Ti30Nb is also the least defected compared to the other alloys under study. However, at this stage, it possesses two distinct grooves about 0.2-0.3 mm wide. In addition, it can be seen that the parameter R_z (average distance between the maximum and minimum points of the surface), has decreased in comparison with the previous step. This indicates that in the process of friction, the indenter gradually sank into the material and uniformly wear the surface. There was no adhesion between Ti30Nb and the counterbody Furthermore, formed debris do not lead to a significantly wear of Ti30Nb surface.

However, the surfaces of Ti and Ti45Nb after 500 cycles have undergone significant changes. Firstly, the profile analysis showed the presence of large surface areas located above the median plane of the material surface (red color), which are, apparently, areas of delamination shift and the result of adhesion of material and counterbody areas. Secondly, the roughness parameters R_a and R_z increased. Moreover, in the case of Ti45Nb, these parameters have more than doubled. In addition, a large number of wear particles were found on the surfaces of these alloys. All this indicates that the adhesion of the surface areas of these materials with areas of the counterbody leads to the appearance of wear particles acting as an abrasive, which in turn leads to a significant decrease in the wear resistance of these materials. It should be noted that there is no straight dependence between microhardness and wear resistance of titanium alloys. Ti45Nb possess higher microhardness, compare with cp-Ti but much less wear resistance.

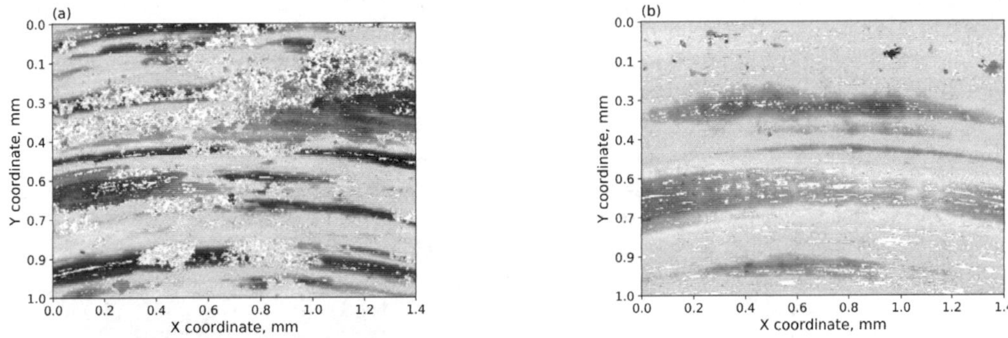

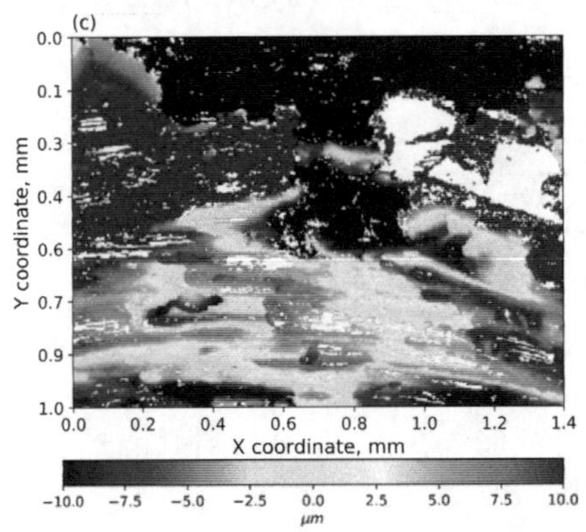

Figure 3. Surface of: a – Ti; b – Ti30Nb; c – Ti45Nb after 500 friction cycles

4. Conclusions

The phase composition of titanium alloys has a significant impact on their frictional behavior. Thus, α and β titanium alloys are subjected to both delamination and adhesion, while titanium with ω phase wear evenly as the indenter sinks. In addition, a significant decrease in wear resistance of α and β titanium also results in the appearance of debris in the process of friction, while their presence does not have a significant effect on titanium with the ω phase. Furthermore, nonstraight relationship between hardness and wear resistance requires *in-situ* experiments of the friction and the study of the structural changes underway.

Acknowledgment
The work was carried out within the framework of the C-19-15 project of a competition among young scientists on the topic "Investigation of the evolution of the structure of surface layers of titanium alloys subjected to simultaneous frictional and corrosive impacts in ex-situ and in-situ modes using synchrotron radiation" in 2019. Materials characterization was carried out at NSTU Materials Research Center.

References

[1] Ozaki T, Matsumoto H, Watanabe S, and Hanada S 2004 Materials Transactions 8 2776–2779
[2] Geetha M, Singh A K, Asokamani R and Gogia A K 2009 Progress in Materials 3 397–425
[3] Niinomi M, Nakai M, and Hied J 2012 Acta 11 3888–3903
[4] Ivanov I V, Thoemmes A and Kashimbetova A A 2018 KEM 769 42–47
[5] Ivanov I V, Thoemmes A, Skiba V Y, A. A. Ruktuev, and I. A. Bataev Met Sci Heat 43 42-46
[6] Niinomi M 2008 J Artif Organs 11 105–110
[7] Niinomi M 2002 Metall and Mat Trans A 33 477–486
[8] Niinomi M 1998 Materials Science and Engineering: A 243 231–236
[9] Kovalevskaya Z G 2016 Metal Working and Material Science 73 34–42
[10] Niinomi M and Nakai M 2011 International journal of biomaterials 2011 836587
[11] Eliades T, Pratsinis H, Kletsas D, Eliades G and Makou M 2004 American Journal of Orthodontics and Dentofacial Orthopedics 125 24–29
[12] Li Y, Wong C, Xiong J, Hodgson P and Wen C 2010 Journal of dental research 89 493–497
[13] Yang X and Hutchinson C R 2016 Acta biomaterialia 42 429–439
[14] Wang K K, Gustavson L J and Dumbleton J H 1996 Medical Applications of Titanium and Its Alloys: The Material and Biological Issues
[15] Li S J et al 2004 Wear 257 869–876
[16] Samuel S, Nag S, Scharf T W and Banerjee R 2008 Materials Science and Engineering: C 28 414–420
[17] Xu L, Xiao S, Tian J, Chen Y 2009 Transactions of nonferrous metals Society of China 19 639-644
[18] Ashiotis G 2015 Journal of applied crystallography 48 510-519
[19] Krishnan H 2013 The bone & joint journal 95 1011-1021
[20] Lee C M, Ju C P, Chern Lin J A 2002 Journal of Oral Rehabilitation 29 314–322
[21] Cai S, Schaffer J E, Ren Y Applied Physics Letters 106 131907

IOP Conf. Series: Materials Science and Engineering **795** (2020) 012010 doi:10.1088/1757-899X/795/1/012010

Neural simulation of ball mill grinding process

A A Zakamaldin and A A Shilin
National Research Tomsk Polytechnic University, 634050 Tomsk, Russia

E-mail: aaz5@tpu.ru

Abstract. This study is aimed at getting simplified model of mill filling technological process of fine crushing in a closed-circuit grinding with screen separation. Optimal and simple model structure are supposed to be used in adaptive predictive control loop. The minor factors that directly affect the mill load indicator are not taken into account, since some of them cannot be directly measured, and other ones affect the process only in the long term. In this paper the athors considered mill filling process identification in the center-discharge ball mill by the method of neural networks (NN). The method includes the identification of the nonlinear process using nonlinear autoregressive with external input (NARX) neural network. The most accurate model was found by varying the structural parameters of the network. The best models were tested in the course of the actual grinding process. The best estimation of the NN model to the real object is obtained with 72.1% match.

1. Introduction

Optimal control remains a complex problem in the mining industry for many years due to various uncertainties in the models of control objects, nonlinearities, changes in parameters and their interdependences [1]. Figure 1 shows the parameters that directly affect to the apatite-nepheline fine crushing process in a closed-circuit grinding with a ball mill and vibrating screens separation.

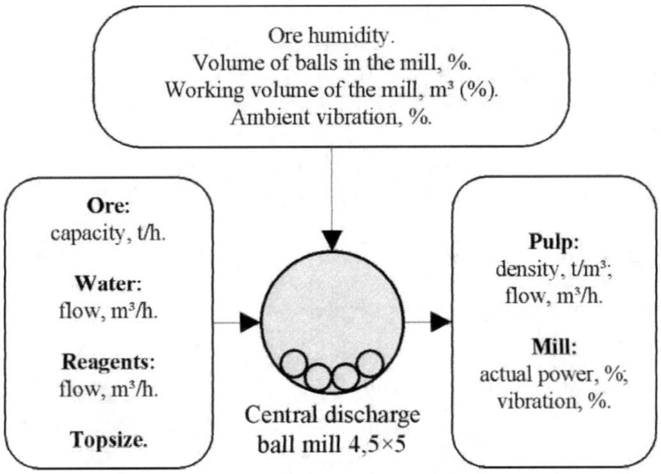

Figure 1. Simplified functional block of the mill.

The flowrates of ore and water are the main material flows entering in the closed grinding cycle. The ore is fed into the mill load by a frequency-driven belt conveyor. The water is supplied at several points of the grinding cycle in order not to clog the chutes and changing the density of the output pulp product to the flotation process. The flowrate of reagents does not affect the processes occurring in the mill. The reagents are fed in the mill to obtain the necessary properties of the output pulp product for flotation. The topsize product returning to the mill from screen separation cannot be measured either qualitatively or quantitatively. The topsize product flow is a dependent parameter on the main input material flows for a stationary process, because the constancy of the technical characteristics of vibrating screens. Other inputs are short-duration perturbations such as ore moisture, ambient vibration and long-term perturbations such as volume of balls in the mill and working volume of the mill [2, 3]. Thus, the number of input parameters indicates the complexity of the object.

The main output parameter of the control object is a load, i.e. material mill filling [4]. Practice shows that the stabilization of the load at the optimum level gives maximum quality indicators of the grinding and a possibility to avoiding the mill overload. The mill overload is achieved if the mill overflows with a material. The overload indicator is the vibration which is measured on the main bearing housing at the discharge chute. This parameter is also called as «noise». The overload adversely affects to the service life of the equipment and the implementation of the grinding plan. In addition, load stabilization is important for the energy efficiency of the entire production process due to the high energy consumption of the process [4].

Modern control systems for grinding include a load stabilization algorithm based on various control approaches [1,6,7]. Domestic algorithms are developed on a cascade PID controller, and still often operate in manual mode or to a limited extent due to the fear of overloading of the mill. PID is stable and efficient, but only around the set of nominal operating points. The permissible overshoot value, determined by the experience of operating the apatite-nepheline fine crushing process control system, is 3% for the control channel. PID approach does not allow achieving this quality of regulation in the mode of the maximum productivity of the mill. Also, advanced control algorithms are actively used and very promising. Model Predictive Control (MPC) is the most popular and successful strategy in non-stationary process control with parameter changes. A special feature of the approach is the using of the process model to calculate the predicted response of the process at future times. The optimal model is the most important part of the MPC strategy. Such a model should cover the key dynamic characteristics of the process and allow calculating predictions [8]. To apply MPC strategy for a particular industrial process, it is necessary to build a custom simulation mathematical model of the object. Thus, to simulate the process of filling the mill with the material, we choose key measuring channel characterizing mill load dynamic: flowrate of ore in the mill / «noise».

2. Description of the method of modelling

The task is to synthesize the optimal structure of the neural model of the control object, to determine the initial parameters of the neural network of the neural model at the current operating point of the control object. The method included the identification of the object with using Neural Network Toolbox 8.4 with the Time Series Tool for the synthesis of custom neural networks with delay lines for input and output signals. The presence of delay lines provides the dynamics of the model, i.e., current output $y(t+1)$ is predicted as a weighted sum of past output values and current and past input:

$$y(t+1) = f(x(t), x(t-1), ..., x(t-d_x), y(t), y(t-1), ..., y(t-d_y)). \quad (1)$$

Equation (1) is called the "regression" equation and d_x, d_y are the numbers of input and output delays required by the autoregressive model.

One important aspect of identification of nonlinear systems is choosing the right time delays for each of the input variables and choosing the number of regressors, i.e., the number of previous samples of each variable that will be considered in the system model at a given moment [9]. The choice of fixed and variable factors is based on the analysis of literary sources [10, 11]. Therefore, the following key factors have changed to build a autoregressive neural network model of the of the control object: d_x, d_y and the number of neurons in the hidden layer. All tested NNs models had the following fixed factors:

- type: recurrent neural network (RNN);
- layers: one input layer, one hidden layer, one output layer;
- transfer function of neurons in hidden layer: tansig;
- transfer function of neurons in output layer: purelin;
- number of neurons in output layer: 1.

The values of the limits and intervals of variation of variable factors:

- number of neurons in hidden layer ranging from 7 to 12 with an interval of 1;
- the number of feedbacks of the input of the neural network (d_x) ranging from 1 to 7 with an interval of 1;
- the number of feedbacks of the output of the neural network (d_y) ranging from 1 to 3 with an interval of 1.

The differences of the RNN model output and control object output are calculated using fit and mean squared error (MSE). Estimations are given by the following equations:

$$fit = \left(1 - \sum_{k}^{n} \frac{(y_k - \hat{y}_k)}{(y_k - \bar{y})}\right) \cdot 100\%, \quad (2)$$

$$MSE = \frac{\sum_{k}^{n} (y_k - \hat{y}_k)^2}{n}. \quad (3)$$

where n – number of training samples, k – model sample time, y_k – real plant output at sample time k, $\hat{y}_k$ – model RNN output at sample time k, $\bar{y}$ – mean real plant output.

All NNs were trained using Levenberg-Marquardtt backpropagation algorithm (trainlm) in 2000 epoch with testing and validation. The data are measured from apatite-nepheline grinding process for training networks. The measurements were provided every second for about 10 hours of mill operation. There were 33555 training input/output samples with 1 second sample time. The input data is the flowrate of ore (t/h) and the output data is the «noise» (%). Trained NNs were checked on training data, on test data and on online data during the grinding process. For online test the laptop is connected to the local computer network of the grinding section automation system. Modbus OPC server is used to establish a connection to the Simulink model and PLC-system. Access to the data "inside" the SCADA-server is carried out according to the standard Modbus rules. We focused on finding the best fitting neural network and in the end compared the best fitting neural network with others Matlab System Identification Toolbox techniques for nonlinear identification: tree partition method, wavelet method, Hammerstein-Wiener model.

3. Results and discussion

It is established that the increase in the number of delays of the output signal by 1 decrease the fit on training data by 2-3 times. That's why further networks were built with one delay of the output signal. In general, 42 neural networks were synthesized varying dx from 1 to 7 and number of neurons from 7 to 12. The learning time increased with the increase in the number of neurons of the hidden layer and dx. Training of some networks stopped on exceeding validation checks. Most of the networks were trained during all 2000 epochs. Performances of all training procedures were less than 0.0001. The

IOP Conf. Series: Materials Science and Engineering **795** (2020) 012010 doi:10.1088/1757-899X/795/1/012010

graphs and results of the online simulation of some models obtained and the output of the real object are presented in figure 2 and in table 1.

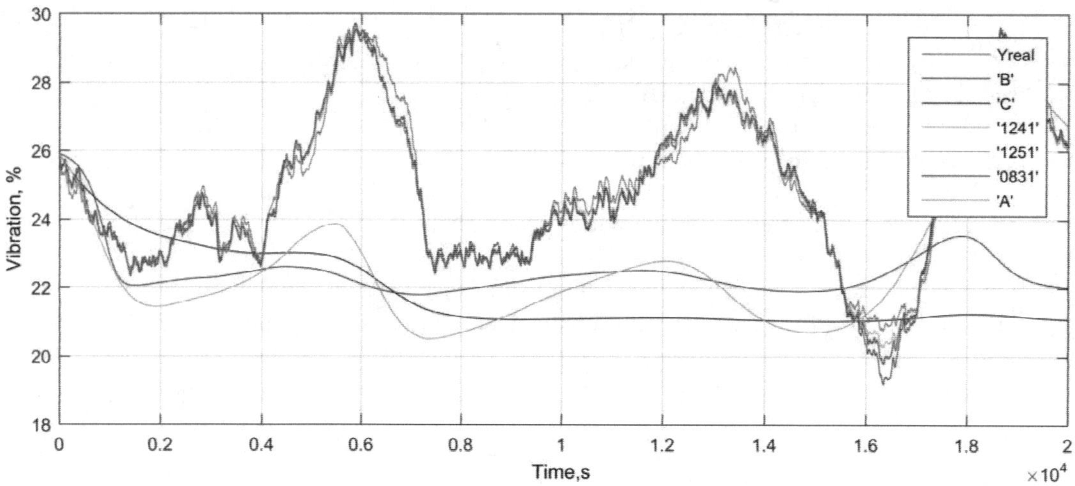

Figure 2. Results of models simulation

Table 1. Results of the simulation.

Model	Training data		Test data		Real data	
	fit, %	MSE	fit, %	MSE	fit, %	MSE
'0831'[a]	92.889	0.0104	65.863	0.2546	72.132	0.2501
'1241'	93.118	0.0098	65.217	0.2643	68.482	0.3200
'1221'	93.770	0.0080	63.107	0.2973	66.428	0.3630
'1271'	94.313	0.0067	61.666	0.3210	64.797	0.3991
'1261'	93.189	0.0096	64.025	0.2827	64.050	0.4163
'1131'	93.244	0.0094	63.856	0.2854	64.027	0.4168
'0941'	95.097	0.0050	64.773	0.2711	63.916	0.4194
'0751'	84.292	0.0508	61.409	0.3253	62.931	0.4426
'0851'	84.292	0.0508	61.407	0.3254	62.643	0.4495
'1251'	84.282	0.0509	61.405	0.3254	62.640	0.4496
'A'	95.473	0.0050	52.848	0.4857	45.429	0.9592
'B'	97.733	0.0039	70.251	0.1933	8.55	2.0971
'C'	97.544	0.0043	67.616	0.2291	5.73	3.2791

[a] Models name decoding: 0831 – 08 - number of neurons in hidden layer; 3 - dx; 1 - dy.

The results in the table 1 are sorted in descending order of fit on the real process data. Three models of other NARX techniques are shown in the end of table 2: 'A' - Hammerstein-Wiener model with 1 output and 1 input (linear transfer function nb = 2, nf = 3, nk = 1, input nonlinearity: pwlinear with 10 units, output nonlinearity: pwlinear with 10 units); 'B' - nonlinearity: wavenet with 25 units, standard regressors: na = 1, nb = 5, nk = 1; 'C' - tree partition method.

An increase in the number of neurons in the hidden layer in the general trend showed an increase in modelling accuracy. The number of input feedbacks of the neural network greater than 3 had an advantage over the others. Comparing with other methods of nonlinear identification, the result of repeating the object on the training sample (33555 samples) and on the test sample (30000 samples) in nonlinear models was higher than the neural models. But on a real process data, these models showed poor results relative to NNs. As in [12], neural networks showed the best results.

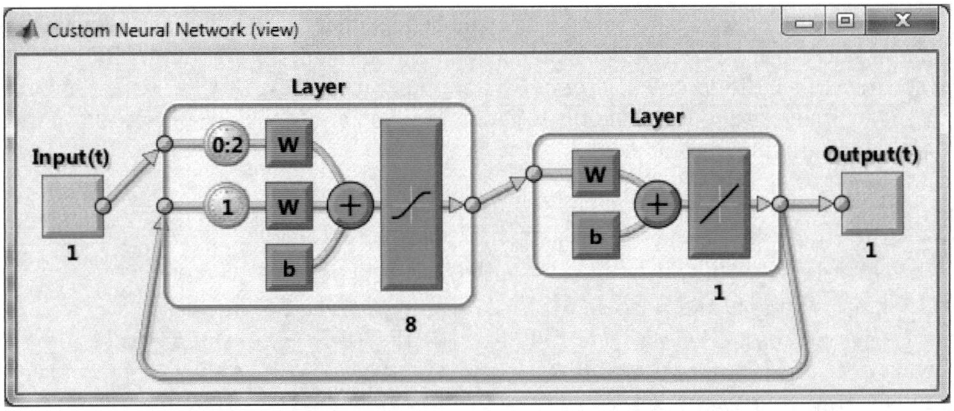

Figure 3. "0831" structure view in Matlab

The following neural network weights for layer 1 (hidden) were obtained for the best fit neural network "0831" architecture (figure 3):

$$w=\begin{bmatrix} w_{1x0} & w_{1x1} & w_{1x2} & w_{1y} & b_1 \\ w_{2x0} & w_{2x1} & w_{2x2} & w_{2y} & b_2 \\ w_{3x0} & w_{3x1} & w_{3x2} & w_{3y} & b_3 \\ w_{4x0} & w_{4x1} & w_{4x2} & w_{4y} & b_4 \\ w_{5x0} & w_{5x1} & w_{5x2} & w_{5y} & b_5 \\ w_{6x0} & w_{6x1} & w_{6x2} & w_{6y} & b_6 \\ w_{7x0} & w_{7x1} & w_{7x2} & w_{7y} & b_7 \\ w_{8x0} & w_{8x1} & w_{8x2} & w_{8y} & b_8 \end{bmatrix} = \begin{bmatrix} 0.2189 & 0.1181 & 0.1113 & 2.9277 & 22.5779 \\ -0.0122 & -0.0297 & 0.0460 & 0.4335 & 26.6599 \\ -0.2558 & 0.4663 & -0.2103 & -4.5690 & -0.9303 \\ 0.0916 & 0.0995 & 0.1035 & 3.6185 & -32.4337 \\ -0.1458 & -0.0604 & -0.0386 & -2.2258 & 32.4675 \\ 0.0398 & 0.0932 & 0.1566 & 1.9836 & -23.2386 \\ -0.1249 & -0.1389 & -0.0293 & -2.2115 & 20.6883 \\ 0.0035 & -0.0198 & 0.0163 & -62.7261 & 0.3475 \end{bmatrix} \quad (4)$$

for layer 2 (output):

$$\begin{aligned} w &= \begin{bmatrix} w_{11} & w_{12} & w_{13} & w_{14} & w_{15} & w_{16} & w_{17} & w_{18} & b_1 \end{bmatrix} = \\ &= \begin{bmatrix} -0.5739 & 0.2819 & -0.0882 & -0.5142 & 0,2982 & -0.7147 & -0.4983 & -0.0158 & 4.0286 \end{bmatrix}. \end{aligned} \quad (5)$$

The obtained neural network model is characteristic of this control object only for the considered steady-state control mode, since disturbances affecting the process are not taken into account in the model. The disturbances need to be taken into account in order to manage quality, so the initial model must be constantly refined. Refining the parameters of the object model in the process of obtaining new data from the worker consists in the repeated parametric identification of the neural network model of the object. Thus, in order to predict the behaviour of an object, it is sufficient to perform online training of a neural network with initial parameters. Re-training should be carried out to achieve the critical deviation of the output predicted by the model from the real output of the object. To do this, it is sufficient to supplement the online algorithm with a recursive parameter estimate to detect deviations of the system operation parameters for a given operating point.

4. Conclusions

Experiments were carried out to search for the optimal model of a ball mill as a control object over the channel "flowrate of ore - noise" using NARX neural networks. We succeeded 72.1% fit of the real control object behaviour with the neural network "0831" architecture. The model can be used to synthesize MPC or advanced nonlinear regulators, including the best fit NN model in the predictive control loop. Also modern methods of neural network modelling are applied in this study and the intuitive algorithm for constructing a process model based on a recurrent neural network is obtained. Undoubtedly, the model obtained by such a method is more complicated from the point of view of solving the MPC problem in comparison with the state-space model. The solution of the MPC problem with the neural model can be obtained by numerical methods with an approximation to the desired simulation accuracy. In addition, it will take more computing resources. However, the flexibility of the neural network approach as a universal approximation function allows us, in spite of the shortcomings, to achieve greater modelling accuracy.

References

[1] Topalov A V and Kaynak O 2004 *J Process Contr* **145** 581-589
[2] Le Roux J D and Craig I K 2017 *IFAC-PapersOnLine* **501** 1169-74
[3] Yu G, Zhou J, Wang Q and Zhao J 2018 *IFAC-PapersOnLine* **5121** 88-93
[4] Tang J, Qiao J, Liu Z, Zhou Z, Yu G and Zhao J 2018 *IFAC-PapersOnLine* **5121** 53-58
[5] Pivnyak G G, Kirichenko V I, Pilov P I and Kirichenko V V 2009 *Gorny Informatsionno-Analiticheskiy Byulleten* **1412** 288
[6] Ulitenko K Y, Sokolov I V and Markin R P 2005 *Tsvetnye Metally* **10** 63
[7] Silva D A, McGarry M F, Rojas AP Gracia R A 2014 *Streamlining the Steps to Optimized Production: Project Process Modelling Advanced Control and Simulator Based Training for Optimized Operation–Case Studies* (Vancouver: CIM Conference)
[8] Tuz A A and Bogatikov V N 2014 *Trudy Kol'skogo nauchnogo tsentra RAN* **5** 24
[9] Bina L, Davidovikj D, Stojanovski G and Stankovski M 2012 *JMEA* **2** 203-207
[10] Khalid M, Yusof R, Joshani M, Selamat H and Joshani M 2014 *Comput-Aided Civ Inf* **293** 221
[11] Eliseev V L 2012 *Razrabotka i issledovanie nejrosetevyh algoritmov upravleniya stacionarnymi i nestacionarnymi ob"ektami [Development and research of neural network control algorithms for stationary and non-stationary objects PhD thesis]* (Moscow: MPEI)
[12] Singh V, Banerjee P K, Tripathy S K, Saxena V K and Venugopal R 2013 *J Powder Metall Min* **2** 2-5

IOP Conf. Series: Materials Science and Engineering **795** (2020) 012011 doi:10.1088/1757-899X/795/1/012011

Information reservation of cyber-physical production of the Industry 4.0

D A Zakoldaev[1], A V Gurjanov[2], A V Shukalov[1], I O Zharinov[1]
[1] Faculty of Information Security and Computer Technologies, Saint Petersburg National Research University of Information Technologies, Mechanics and Optics, 49, Kronverksky Av., Saint Petersburg, 197101, Russia
[2] Department of Aviation Instrumentation Systems and Training Systems, Saint Petersburg State University of Aerospace Instrumentation, 67, Bolshaya Morskaia str., Saint Petersburg, 190000, Russia

E-mail: mpbva@mail.ru

Abstract. An actual task has been studied to organize informative reserving of the production data and link channels in the Industry 4.0 smart factory. Informative reserving principles in a digital company are done by duplicating informative link channels into a single company communication environment. Connection channels duplicating creates the production division informative redundancy for the technological operations completion results data. The smart factory informative reserving increase company functionality reliability and provides functional and informative smart factory functionality. There are schemes of cyber and physical production components informative reserving and types of connection channels are defined which are used as primary and reserve lines of the information transmission. There are ways of informative reserving based on the technologies of Bluetooth, Ethernet and Internet of Things (IoT). Main way of cyber and physical production informative reserving is done in system level and is duplicating of the IoT wireless channel with a wired channel of the Ethernet connection and transmitting of duplicating messages. Auxiliary way of cyber and physical production informative reserving is done in local level of the production data exchange among cyber and physical systems with a Bluetooth channel.

1. Introduction

One of the most important property of the Industry 4.0 [1, 2] production company is the company informative transparency. The smart factory informative transparency is the personnel possibility to monitor the item manufacturing technological operation completion state at any stage of the technical cycle [3].

To realize the company informative transparency they must create the company communication infrastructure which grants the operator access to the production data which characterizes item manufacturing technological processes [4]. The communication infrastructure unites the cyber and physical systems, cloud resources, operator access means to the company industrial net and other production components [5, 6].

Computerized communication infrastructure makes in automatic way some smart factory item manufacturing processes but the system access control to the production components has some risks of the company informative security [7, 8]. So along the task of the CPS machine reserving in a cyber and physical production there are also tasks of the company informative security including the tasks of

IOP Conf. Series: Materials Science and Engineering **795** (2020) 012011 doi:10.1088/1757-899X/795/1/012011

the resources and production data informative reserving [9, 10].

The informative reserving is applied in production if a technological line has a great number of cyber and physical systems (CPS) to generate homogeneous production data but with different physical principles. The sources of such production data are measuring detectors of sensor net and controllers connected to the company communication environment [11].

The CPS production data informative reserving principle is an intentional redundancy of connection channels, data processing centers and information sources for the production division stable conditions of working if some components are failed [12]. Informative reserving is the ability of cloud applications to process a restricted volume of production data without any loss of accuracy and cyber and physical systems [13] control algorithms resistivity. Cyber and physical production informative reserving can be done in two levels of control:

- in the system level of production components interaction, cloud services and applications;
- in the local level of some cyber and physical systems interaction.

2. Informative reserving in the system level

Cyber and physical production informative reserving scheme in the system level is given in figure 1.

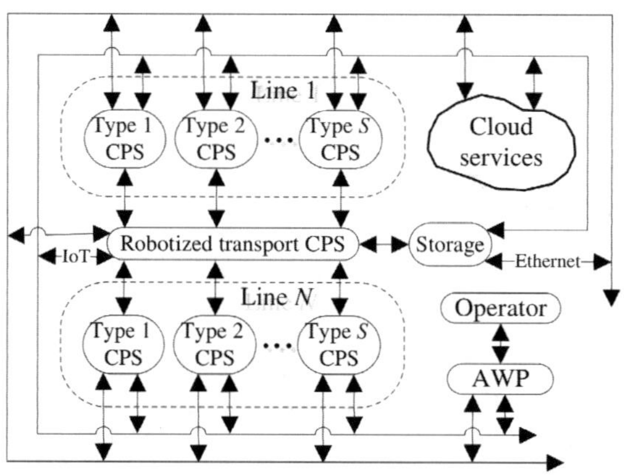

Figure 1. The Industry 4.0 cyber and physical production informative reserving scheme in the system level.

Cyber and physical production informative reserving in the system level is done by duplicating of the industrial Internet of Things (IoT) wireless communication environment with the wired channels of Ethernet connection. If transmitter and receiver fail which are parts of a CPS and provide the connection with the channels of IoT the information data exchange for this CPS will be conducted through a channel of Ethernet. Both connection channels provide CPS connection in parallel to the cloud services and applications. This way of cyber and physical production informative reserving is to transmit production data to the cloud centers of processing which makes available the BigData technology for cloud services and applications.

The informative reserving of cyber and physical production in the system level is the highly-reliable way to increase the functional safety of the Industry 4.0 smart factory. But anyway if the production data processing center analytical component fails the functionality of the entire company production division can be paralyzed.

3. Informative reserving in the local level

An effective way to increase the cyber and physical production reliability is an informative reserving in the local level of some CPSs interaction. The cyber and physical production informative reserving scheme in the local level is given in figure 2.

Cyber and physical production informative reserving in the local level based on CPS interaction in wireless link channels with the protocols of Bluetooth, a channel duplicating the Industrial Internet of Things. The link channel of Bluetooth helps to organize informative interaction of neighboring CPSs installed in production with a short distance from each other.

CPS interaction with channels of Bluetooth must be viewed as a mean of control of the product being manufactured when a CPS may check the result of the done operation to provide the next processing of details. When transmitting and receiving pieces of CPS IoT information exchange fail the CPS interaction will continue with a channel of Bluetooth. A feature of the Bluetooth channel is an ability to transmit data for a short distance that the CPS information exchange could be done in pairs. Within a technological section Bluetooth protocols interaction of CPSs is done in chain with neighboring CPSs in the information exchange radio channel.

Exceptionally if a CPS channel of IoT and Ethernet fails such a CPS will access cloud services and applications with Bluetooth and with working means of data transmitting to the neighboring CPS.

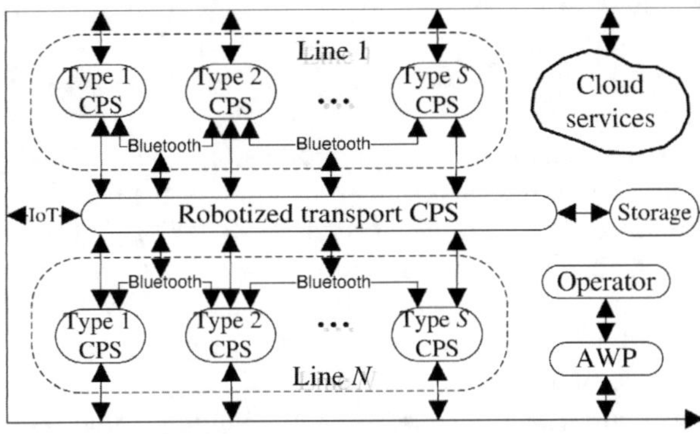

Figure 2. Cyber and physical production informative reserving scheme in the local level.

Difference in the pass-through ability of the link channels of Ethernet, IoT and Bluetooth may create a production realization with informative redundancy and time redundancy. The temporal redundancy is a way of production data reserving when technological operation completion state information is transmitted to the control computerized system several times (For highly-productive link channels). Multiple transmission of production information increases the data accuracy being circulated in a smart factory. Production data integrity and accuracy is the most important property of computerized control system information.

A feature of informative reserving in cyber and physical production is the CPS digital twins in the company virtual environment organized in the level of cloud services and applications. CPS digital twin has the full information similarity to the original (CPS physical device) which is for calculation processes organization and CPS control processes in cyber and physical production in the company virtual environment. The company virtual environment is for imitation modelling of technological processes which happen in CPS real working chambers which together with CPS detector data provides the increased accuracy of the assembly units manufacturing with the methods of complex information processing.

Together with cyber and physical technological equipment digital twins the smart factory cloud environment contains the parts digital twins being manufactured in production. Parts digital twins are an electronic technical documentation represented in the level of mathematical, program, construction and technological models which describe item different properties. The part digital twin is a model being developed in time which improvement depends on the data received from the exploitation of the manufactured before items. So cyber and physical production is an informative data connection of return which characterizes quality and specifications of the items being manufactured. The item digital twin models are used in calculations of the items being manufactured parameters being realized in the smart factory virtual environment. The application of item digital twins duplicating the physical samples of the items being manufactured is a special form of the Industry 4.0 cyber and physical production informative reserving.

4. Conclusion

Cyber and physical production projection with reserved components is a science and practical task which solution lies with the methods of the linear programming. Methods of the linear programming (simplex-method, artificial basis methods and other) helps to optimize the maximum number of CPS placement including duplicating CPSs in a particular site of a smart factory. Initial data for automatic production projection with the methods of linear programming are the lengths of a smart factory sides (production division), dimensions of all CPSs and distances between each CPS pair and also distances among CPSs and external elements of the company.

Digital informative technologies implementation in smart factories creates the necessary conditions of the electronic documentation inside the company and production company interaction with its agents of production cooperation. Standardizing and creating types of program interfaces and environments to unite the profile smart factories into distributed virtual companies (virtual factories) for the item mass production.

Electronic documentation acceptability (digital models) in the technological cycle «projection-production» helps to integrate the virtual factories projection companies (digital factories) to form the finished structure of the Industry 4.0 in the format of a factory of the full cycle of the item manufacturing. Information spaces of separate companies unification into a single virtual factory creates the informative redundancy of project and production data which based on the BigData algorithms must be converted to informatively reserved project and production data to increase reliability and informative safety of a digital company functioning.

Technological operations completion processes documentation in the Industry 4.0 smart factory is done in virtual environment company servers. Each cyber and physical systems including the transport CPS and production control computerized system during the functionality of scenarios which content is written in special files in the digital cloud. Scenario files analysis is done by the operator in the smart factory running in field works or in the stage of malfunctioning search, which are consequences of cyber and physical equipment failures.

References

[1] Rennung F, Luminosu C T, Draghici A 2016 *Procedia social and behavioral sciences* **221** 372-377
[2] Brik B, Bettayeb B, Sahnoun M and Duval F 2019 *Procedia computer science* **151** 667-674
[3] Bogatyrev V A, Parshutina S A, Poptcova N A and Bogatyrev A V 2016 *Communications in computer and information science* **678** 337-348
[4] Zhong R Y, Xu X, Klotz E and Newman S T 2017 *Engineering* **3** 616-630
[5] Schumacher A, Nemeth T, Sihn W 2019 *Procedia CIRP* **79** 409-414
[6] Shihundla T B, Mpofu K, Adenuga O T 2019 *Procedia CIRP* **83** 8-13
[7] Sallati C, Bertazzi J A, Schutzer K 2019 *Procedia CIRP* **84** 203-208
[8] Gurjanov A V, Zakoldaev D A, Shukalov A V and Zharinov I O 2018 *IOP Conference Series: Materials Science and Engineering* **327** 022110
[9] Zakoldaev D A, Shukalov A V, Zharinov I O and Zharinov O O 2018 *Journal of Physics: Conference Series* **1015** 052033
[10] Peruzzini M, Grandi F, Pellicciari M 2017 *Procedia manufacturing* **11** 806-813
[11] Benesova A, Tupa J 2017 *Procedia manufacturing* **11** 2195-2202
[12] Suarez F-M, Marcos M, Peralta M E and Aguayo F 2017 *Procedia manufacturing* **13** 1229-36
[13] Mantravadi S, Moller C 2019 *Procedia manufacturing* **30** 588-595
[14] Kuo C-C, Shyu J Z, Ding K 2019 *Global transitions* **1** 3-14

IOP Conf. Series: Materials Science and Engineering **795** (2020) 012012 doi:10.1088/1757-899X/795/1/012012

Estimation of Stress-Deformed State of the Cutting Insert of a Disk Mill Using Computer-Aided Engineering

V G Shalamov, S D Smetanin and I S Boldyrev
Federal State Autonomous Educational Institution of Higher Education "South Ural State University (national research university)", 76 Lenin prospekt, Chelyabinsk 454080, Russia

E-mail: smetaninsd@susu.ru

Abstract. The efficiency of the assembly tool is largely determined by the cutting tool insert installation scheme, as well as by the accuracy of manufacturing mating surfaces and loading conditions. When milling the normal types of destruction of the cutting blade are chipping and micro-dyeing. In this regard, in order to improve the reliability of a tool, it is desirable to provide the most uniform distribution of stresses at the initial time. In the article, the stressed-deformed condition of a cutting blade of a disk mill is considered taking into account the schemes of the cutting tool insert basing, the clamping force and the external cutting force. The stages of solving the problem of determining the stressed-deformed condition design of a disk mill are shown. The ratios of the force of clamping of the cutting tool insert and external force are determined. The joint effect on the stress state of the cutting tool insert of the clamping force and external force is analyzed.

1. Introduction

Cutters are a universal widespread tool that provides high performance machining. Depending on the position of the axis relative to the machined surface, milling cutters are disc, end and finger, and in construction - integral and prefabricated. When using the cutting part of a hard alloy as a material, only small-sized tools are manufactured. Most of the carbide milling structures are prefabricated, that increase durability and reduce tooling costs. A wide range of inserts causes a variety of both the conceptual diagrams of their installation in the housing, and the variants of realizations in concrete structures.

The main reasons for the deterioration of cutting insert are: high power and temperature loads, low strength and wear resistance of hard alloy, non-optimal structural and geometric parameters. This makes it necessary to determine the stress-strain state and strength, i.e. in the study of the distribution of dangerous stresses, leading to destruction. Many works have been devoted to the study of the stress-deformed state of the cutting edge, but they practically do not consider the design of the tool as a whole, which leads to a lack of understanding of the behavior of its various components. Therefore, it will be useful from a scientific and practical point of view, the study and coordination of stress-deformed state on the surfaces of the components of the mill design.

Designing a new design requires considerable time and money. Simulation is used to reduce them. Modeling is an effective and now universally applicable method of creating any objects, processes and phenomena. The main purpose of the model is to make it possible to predict the behavior of the projected object with permissible changes in its structure and parameters (comparative analysis of alternative design options) and environmental conditions (research of the object in various situations). The purpose of modeling the stressed-deformed state of the milling design is to synthesize the amount of tightening of the insert when a cutting force is applied to the cutting element.

2. Methodology

The efficiency of the assembly tool is largely determined by the cutting insert installation scheme, the accuracy of manufacturing the mating surfaces and loading conditions. Production experience has shown that, when milling, the destruction of the insert is fragile, and the widespread types of destruction of the cutting blade are chipped and micro-dyeing. In this regard, in order to increase the reliability of the tool, it is desirable to provide the most uniform distribution of stresses in the cutting blade at the initial time. In addition, practical experience shows that the operability of cutters with indexable inserts essentially depends on the adopted scheme of basing, the binding force and the joint influence of the binding force and external force.

The finite element method (FEM) is widely used in the study of cutting processes. The most common case is orthogonal cutting. The paper [1] presents 2D thermo-mechanical FEM model of orthogonal cutting. The stresses in the primary and secondary deformation zone are simulated. Concluded that the variation of friction coefficient affects thrust force component and temperatures more than it affects cutting force.

The greatest amount of research is devoted to turning [2-10]. The authors are explored stress, strain, temperature, chip formation mechanisms and other characteristics of turning process. Several studies show the modeling of the broaching process [11, 12]. The FEM was applied [13] for understanding and predicting the residual stress values during ball-burnishing, allowing to determine the optimal process parameters. Modeling and simulation of the laser cutting process [14] using the analysis of heat transfer and general energy balance analysis made it possible to determine of temperature and stress distribution along with heat-affected zone. In the research of the deep twist drilling process, the authors [15] simulated the light drilled structure and the conventional welded structure and analyzed their behavior under compression loads. The von Mises stresses and the displacement of the models are determined. A lot of research has been also devoted to modeling the milling process. With the help of finite element analysis, the effect of the cutting speed on the cutting force, the temperature, width and curl radius of chip, and the wearing depth of tool are investigated over the milling process [1, 16-20].

In all these works, the finite element method is used to model the interaction of the tool with the workpiece. The integrated model proposed by the authors [21] takes into account the clamping preloads, machining forces, locator–workpiece contact interaction, fixture compliances, machine table stiffness, and forced vibrations. The FEM in study [22] creates spindle, holder and tool models and is used to simulate the contact conditions of tool and workpiece as well as spindle and tool holder and contact of tool-workpiece. In the paper [23], the authors illustrate the analysis of tooltip positioning error along the three coordinate axis in a CNC machine. This study takes into account only static loads, neglects the dynamic effect during milling, and assumes that the tool and the workpiece deform to their static equilibrium positions at any milling instant.

At the same time, in open access there is no information on modeling and optimization of the design of mills with the finite element modelling.

In the designs of prefabricated milling cutters, a circular form is often used. The round insert has the advantage associated with the lack of corner areas, which are stress concentrators. Besides, when turning the insert with worn cutting edge around axis, it is possible to use practically the entire perimeter of the cutting edge, thus increasing tool life and reducing costs. However, the circular cutting edge determines the various deformations of the cutting layer associated with the direction of the outgoing chip removal, causing a different amount of cutting force. Therefore, modeling the design of cutters with round inserts requires consideration of specific features.

For the analysis of stressed-deformed state, a 3D model of a disk milling cutter (Figure 1) consisting of a housing 1, a cassette 2 with a pressed pin 4, a cutter insert 3, a ring 5, a washer 6 and a screw 7 is created. When the screw 7 is screwed, the cassette 2 moves radially, thanks to which the insert 3 is pressed against the base surface of the housing 1. This design corresponds to the fastening system of the plate, designated according to the ISO 518 standard by the letter P: the basing of the insert in the tool body along the support and side surfaces with a pin clamped through the central hole. The method provides a guaranteed pressure against the surface of the body, high accuracy of positioning of the main cutting edge, good chip removal, coincidence in the direction of cutting forces and fastening.

The cutter can cut through the grooves with a depth greater than the radius of the insert, which imposes a limit on the thickness of the shell. In this paper, we consider a round insert with a clearance angle. At the base of the conical lateral surface of the insert, two circuits are possible in the cutter body (Figure 2): in the first scheme (Figure 2a), the contact of the conical surface of the insert with the body is carried out by a smaller cone diameter, and in the second scheme (Figure 2b) - larger. Evaluation of stressed-deformed state in both cases requires modeling of two versions of the housing, the remaining elements of the design remain unchanged.

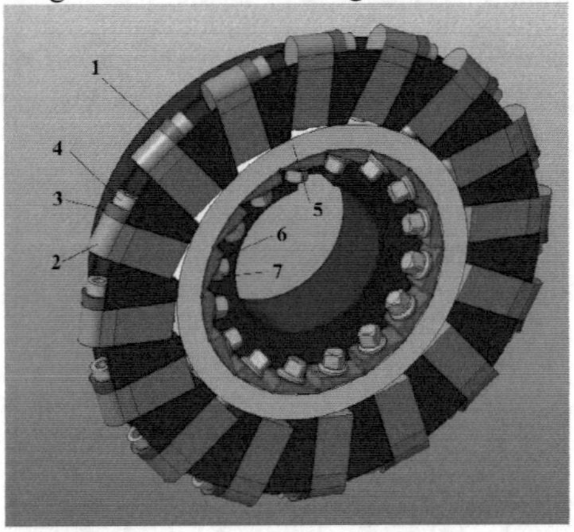

Figure 1. 3D model of milling cutter.

The developed design scheme (Figure 2c) takes into account the application scheme of the components of the cutting force and subsequent modeling (clearances $\Delta 1$ and $\Delta 2$). Due to the symmetry of the cutting edge, the axial component of the cutting force (in the X-axis direction) is not taken into account, but only the radial Py and the tangential Pz components are taken into account. Also, an assumption has been made that the component cutting forces are distributed evenly along the entire cutting edge (in a half circle). The clamping force of the Pc is directed parallel to the cassette's reference plane. Pressing the pin into the holder allows, when performing calculations, to accept them as a single component (cassette).

The methodology for calculating the stressed-deformed state includes three successive steps that correspond to the actual construction of the mill.

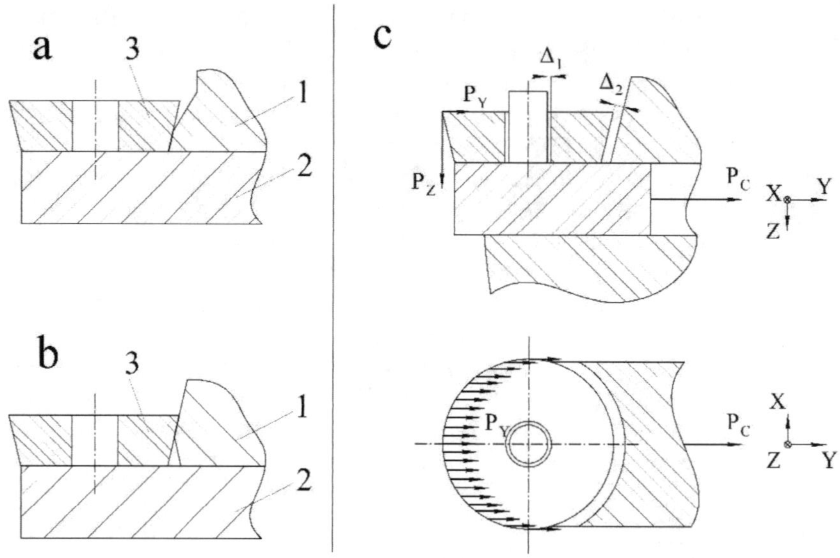

Figure 2. Basing schemes of insert: (a) first scheme; (b) second scheme; (c) schematic view of a simplified model of a disk mill.

Step 1 - ensuring the specified relative arrangement of the structural elements. The housing is rigidly fixed, i.e. there is no its movement along the coordinate axes. At this point, there are gaps between the assembly components: Δ1 - between the pin and the insert and Δ2 - between the insert and the housing. When performing calculations at this step, the absence of interference (interference) between the components whose criterion is the magnitude of the stresses is checked (theoretically it should be zero).

Before performing the calculations, it is necessary to spatially sample the calculated area, i.e. replacement of the solid model by a net of finite elements (FE). This requires a reasonable choice of the type and size of the FE. The type of FE is due to the studied properties of the object, its shape and dimensions. Among the volume elements, the FE in the form of a cube and a regular tetrahedron provides the best convergence conditions for the solution. We take as the FE the type SOLID185, used for three-dimensional modeling of volumetric constructions. It is determined by 8 nodes (cube) in 3 degrees of freedom in each. The element has the properties of plasticity, hyperelasticity, hardening, creep, deformations and displacements. It can take the form of a 6 (prism), 5 (pyramid) or 4 (tetrahedron) node element, if it is necessary to coordinate the mesh with the partition. The size of the FE is chosen on the basis of a trade-off between sampling errors and the required computing power resources. When choosing the size of FE for prefabricated structures, it is desirable to observe the proportionality of different components. Therefore, the size of the FE for the housing is taken equal to 1 mm, and for cassette and insert - 0.5 mm.

The model should include the minimum necessary part of the design, allowing to obtain an adequate solution at the minimum cost of the calculation time. Since the design is axisymmetric, to create a finite element mesh and accelerate the calculation in the computer-aided engineering system, it is sufficient to have one sector of the enclosure containing one cassette with the insert (Figure 3).

The initial data include the properties of materials: the Young's modulus E and the Poisson's ratio μ. As the material of the housing and the cassette, structural steel is adopted, and the insert is a hard alloy.

The boundary conditions are as follows: rigid fixation of the housing in all coordinates; limit the movement of the cassette along the X and Z axes; no gaps between the cassette and the housing.

We accept contact interactions with a friction coefficient of 0.25: between the flat surfaces of the housing of mill and the cassette; between the conical surface of the insert and the housing; between the lower surface of the insert and the plane of cassette; between the hole in the insert and the cylindrical surface of the pin.

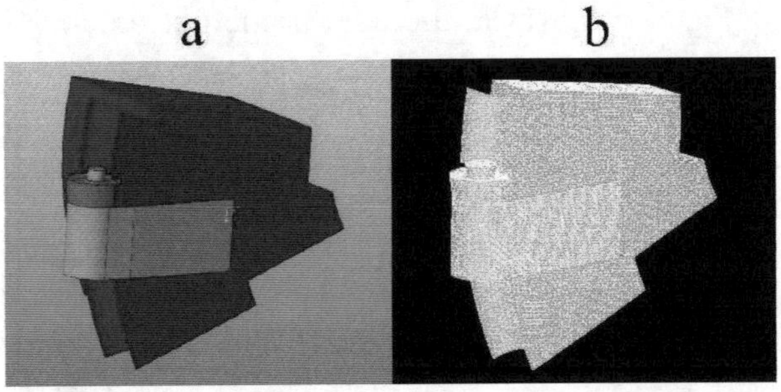

Figure 3. (a) the mill sector; (b) its representation by a finite element mesh.

Step 2 - insert holder tightening. It occurs when assembling the mill before work or changing of worn insert. To do this, the Pc clamping force is applied to the end face of the cassette in the direction of the Y axis. As a result, the screw presses the insert to the base surface in the housing, creating a certain interference. The corresponding stresses appear on the components of the structure.

Step 3 – modeling the cutting process. A uniformly distributed cutting force is applied to the cutting edge of the insert. The value of the tangential component of the cutting force is assumed equal to 500; 1000; 1500; 2000; 2500 N, and radial - equal to half of the tangential component. There is a redistribution of stresses resulting from the clamping of the cassette.

The resulting tension in the cassette is concentrated on the cylindrical surface of the pin (Figure 4a). As a result of the application of the components of the cutting force, the insert is clamped to the support plane of the cassette, which causes the appearance of stresses on it (Figure 4b). After the application of the cutting force, the redistribution of stresses between the pin and the reference plane occurs: a decrease on the pin and an increase in the plane (Figure 4c).

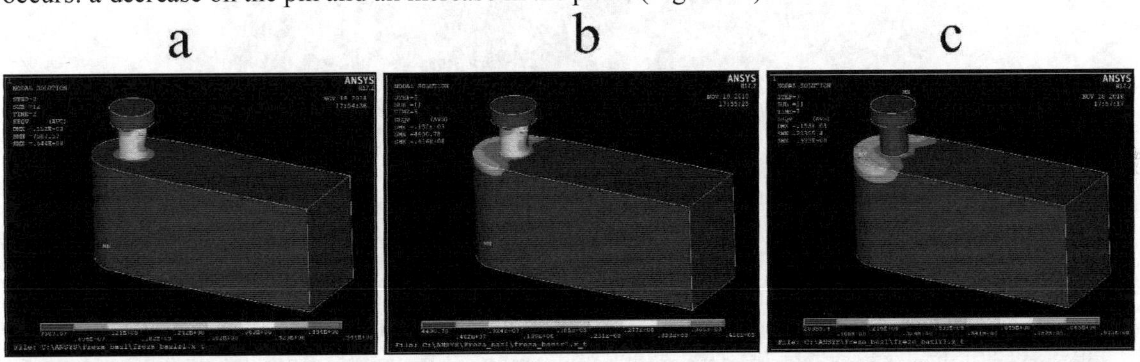

Figure 4. Stress change with increasing cutting force: (a) without cutting force; (b) with minimum cutting force; (c) with maximum cutting force.

In general, the magnitude of the stresses on the surface of the pin and the support surface of the cassette are different. However, with a certain clamping force, it is possible to determine such a cutting force at which the stresses on the pin 2 and the support plane 1 are aligned (Figure 5a). This condition can be reflected graphically: the interrelation of the clumping force and the cutting force (Figure 5b). Therefore, for a given cutting force, it is possible to determine the clamping force, which ensures equal stresses on the holder and the pin. If necessary, it is possible to ensure the equal strength of these elements, taking into account their geometrical parameters and the type of loading. It can be noted that with insufficient clumping force, the calculation fails. The excessive clamping force leads to stress in the structural elements of the cutter above the conventional yield strength of structural steel

IOP Conf. Series: Materials Science and Engineering **795** (2020) 012012 doi:10.1088/1757-899X/795/1/012012

(Figure 6a), which leads to plastic deformation and violation of proper functioning. Also, with an increase the clumping force the stress at the point of contact of the insert, they become higher than the stresses at the cutting edge (Figure 6b), which can cause brittle fracture of the insert not in the cutting zone.

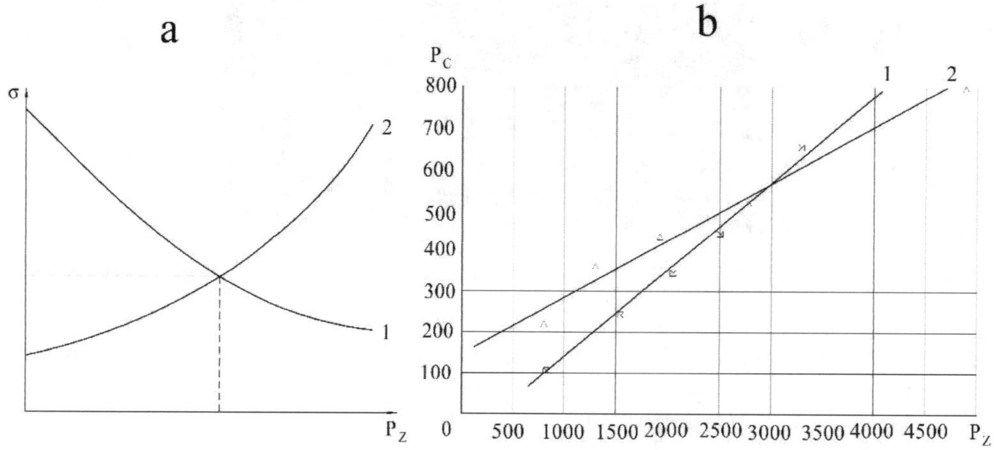

Figure 5. (a) the relationship of stress and cutting force; (b) interconnection of cutting and clumping forces.

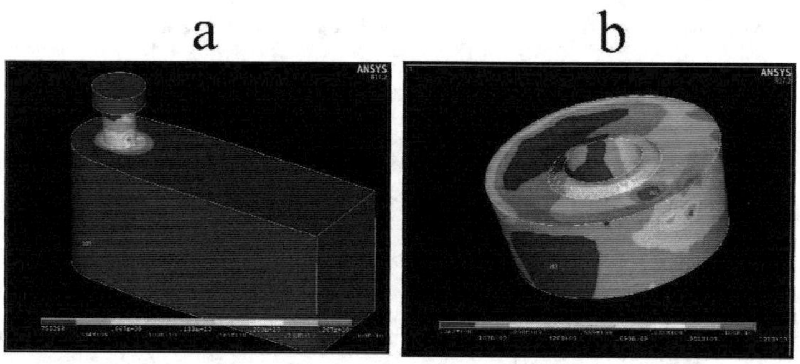

Figure 6. Von-Mises stress in the elements of the cutter design: (a) maximum stresses $0.3*10^{10}$ Pa; (b) stresses in the insert $0.121*10^{10}$ Pa.

An analysis of the graphs of the relationship of cutting and clumping forces shows that the basing scheme affects the magnitude of the required clumping force. With small cutting forces (up to 3000N) basing according to the first scheme requires the application of a smaller clamping force. In addition, this scheme provides a closer to the linear relationship of forces.

3. Conclusions

In the article, the stressed-deformed state of a cutting blade of a disk mill is considered taking into account the insert scheme of basing, clamping force and cutting force.

If the clamping force of the cassette is insufficient after loading, even with the minimum cutting force, infinitely large displacements of the insert occur, leading to a failure in the calculation. In practice, this means dropping the insert from the housing of mill.

The excess clamping force causes the occurrence of stresses above the conditional yield strength of the structural materials of the components. In practice, this will cause plastic deformation of proper functioning. The brittle destruction of insert may not occur in the cutting zone.

The clamping force can be assigned depending on the expected cutting force, which will ensure in the process of work the equality of the stresses on the holder and the pin. The basing of the first scheme with low cutting forces is preferable from the point of view of lower stresses in the holder and pin.

Continuation of the research can be done by comparing the stressed-deformed state with other types of disk milling constructions, examining various insert materials, changing the insert geometry and the direction of cutting force, taking into account temperature deformations.

Acknowledgments
South Ural State University is grateful for financial support of the Ministry of Education and Science of the Russian Federation (grant No 9.5589.2017/8.9).

References
[1] Yuan Y, Jing X, Ehmann K F et al 2018 Modeling of cutting forces in micro end-milling *J. of Manufacturing Processes* **31** 844–858
[2] Markopoulos A P, Vaxevanidis N M and Manolakos D E 2015 Friction modeling in finite element simulation of orthogonal cutting *Tribology in Industry* **37** 440-448
[3] Yan H, Hua J and Shivpuri R 2005 Numerical simulation of finish hard turning for AISI H13 die steel *Science and Technology of Adv. Materials* **6** 540–547
[4] Bassett E 2014 Belastungsspezifische Auslegung und Herstellung von Schneidkanten fu¨r Drehwerkzeuge (Hannover)
[5] Ma Y, Zhang J, Feng P et al 2018 Study on the evolution of residual stress in successive machining process *Int. J. Adv. Manuf. Technol.* **96** 1025-34
[6] Ortiz-de-Zarate G, Madariaga A, Garay A et al 2018 Experimental and FEM analysis of surface integrity when broaching Ti64 *Procedia CIPR* vol 71 (Elsevier) pp 466-471
[7] Peirovi S, Pourasghar M, Nejad A F et al 2017 A study on the different finite element approaches for laser cutting of aluminum alloy sheet *Int. J. Adv. Manuf. Technol.* **1** 1-15
[8] Lopez de Lacalle L N, Fernandez A, Olvera D. et al 2011 Monitoring deep twist drilling for a rapid manufacturing of light high-strength parts *Mechanical systems and signal processing* **25** 2745–52
[9] Rai J K and Xirouchakis P 2008 Finite element method based machining simulation environment for analyzing part errors induced during milling of thin-walled components *Int. J. of Machine Tools and Manufacture* **48** 629–643
[10] Pour M and Ghorbani H 2017 Improving FEM model of low immersion milling process using multi-objective optimization of tool elastic support dynamic properties *Int. J. Adv. Manuf. Technol.* **92** 2279–97
[11] Rodríguez A, López de Lacalle L N, Celaya A et al 2012 Surface improvement of shafts by the deep ball-burnishing technique *Surface and Coatings Technology* **206** 2817-24
[12] Zhang P and Wang Y 2018 A study on chip and microstructure of 7055 aluminum alloy's 3D HSC based on FEM and experiment *Vacuum* **152** 205-213
[13] Bermudo C, Sevilla L, Martín F et al 2016 Study of the Tool Geometry Influence in Indentation for the Analysis and Validation of the New Modular Upper Bound Technique *Applied Sciences* **6** 1-16
[14] Xiong Y, Wang W, Jiang R et al 2018 Mechanisms and FEM Simulation of Chip Formation in Orthogonal Cutting In-Situ TiB2/7050Al MMC *Materials* **4** 1-19
[15] Biermann D, Menzel A, Bartel T et al 2011 Experimental and computational investigation of machining processes for functionally graded materials *Procedia Engineering* **19** 22–27
[16] Xie L-J, Schmidt C, Biesinger F et al 2009 Wear progress prediction of carbide tool in turning of AISI1045 by using FEM *Proc. of CIST2008 & ITS-IFToMM2008* (Beijing: Elsevier) pp 372-375

[17] Krajinović I, Daves W, Tkadletz M et al 2016 Finite element study of the influence of hard coatings on hard metal tool loading during milling *Surface and Coatings Technology* **304** 134-141
[18] Liao Y G and Hu S J 2001 An Integrated Model of a Fixture–Workpiece System for Surface Quality Prediction *Int. J. Adv. Manuf. Technol.* **17** 810–818
[19] Zhang Q, Zhang S and Li J 2017 Three dimensional finite element simulation of cutting forces and cutting temperature in hard milling of AISI H13 steel *Procedia Manufacturing* **10** 37–47
[20] Yanda H, Ghani J A, Rizal M et al 2015 Performance of uncoated and coated carbide tools in turning FCD700 using FEM simulation *Int. J. Simul. Model.* **14** 416-425
[21] Klocke F, Lung D and Buchkremer S 2013 Inverse identification of the constitutive equation of Inconel 718 and AISI 1045 from FE machining simulations *Procedia CIRP* **8** 212–217
[22] Pour M and Ghorbani H 2017 Improving FEM model of low immersion milling process using multi-objective optimization of tool elastic support dynamic properties *Int. J. Adv. Manuf. Technol.* **92** 2279–97
[23] Afkhamifar A, Antonelli D and Chiabert P 2016 Variational analysis for CNC milling process *Procedia CIRP* **43** 118-123

IOP Conf. Series: Materials Science and Engineering **795** (2020) 012013 doi:10.1088/1757-899X/795/1/012013

Analysis of Pressure and Dynamic Tests of the Blocked Wheel Pair

V V Shiler[1], A V Shiler[1], I I Galiev[1] and G I Petrov[2]

[1]Omsk State Transport University (OSTU), 35, Marks Avenue, 644046 Omsk, Russia
[2]Russian University of Transport, 9-9, Obraztsova, 127994, Moscow, Russia

E-mail: shiler_alex@inbox.ru

Abstract. The paper describes a modular design of a wheelset. Three prototypes of wheelsets and track were manufactured using the π theorem. Full-scale wheelsets for a cargo car truck were also manufactured. The results compare the movement dynamics of conventional and modular wheel-sets mounted on cargo car trucks.

1. Introduction

The current direction in improving the efficiency of rail transportation is to reduce the stress state during the interaction of elements in the mechanical system "wheelset - track". World experience in the operation of railways shows that the process of dynamic interaction of rolling stock and track is the key in solving the problems of railway traffic. Therefore, technical solutions aimed at reducing the dynamic interaction and tension of the elements of the mechanical system “wheelset - rail-sleeper grid” should be used along with traditional methods to increase the parameters of reliability and durability. Such approaches are more effective and less costly.

2. Experimental section

The mock-up tests involved modular wheelsets with the design presented in Figure 1 [1]. The wheelset contains two independently rotating wheels 3 and 5. The guiding wheel 3 rigidly sits on axis 8, which can rotate due to box bearings 9. The guiding wheel 3 has a flange and receives horizontal forces from the side surface of rails. The carrying wheel contains wheel center 5 and is mounted through pair of bearings 13 on the wheelset axis 8 and transfers via elastic gasket 11 the vertical load of the rolling stock weight to flange 1 which rolls along the roll surface of the rail. Therefore, the modular design implements independent rotation of the roll surfaces of the wheels that during movement contact with the track rail top surfaces. Due to rubber gasket, the elastic flange is cushioned in relation to the wheel center, which enables elastic suppression of oscillations caused by irregularities on the roll surfaces of wheels and rails.

The physical model of the wheelset is based on the theory of similarity of mechanical systems, which is a part of the fundamental theorem – so-called π theorem – analyzing the dimensions in its classical formulation [2, 3]. For the systems (original and model) to be mechanically similar, the requirements of geometrical, material and kinematic similarity should be met.

IOP Conf. Series: Materials Science and Engineering **795** (2020) 012013 doi:10.1088/1757-899X/795/1/012013

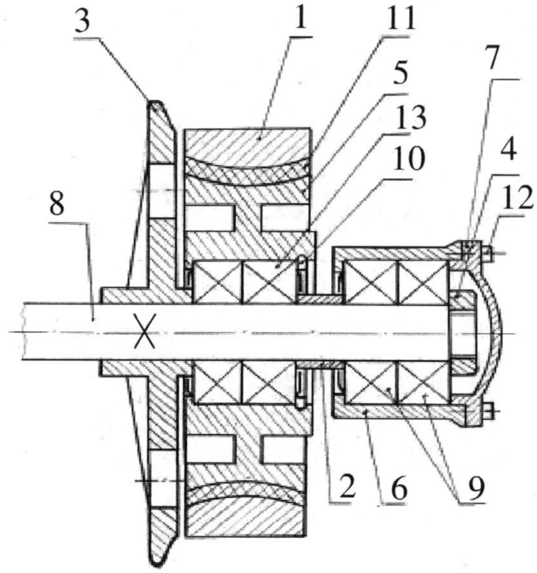

Figure 1. Novel design of a wheelset wheel of a rail vehicle.

Three models of track, car trucks and wheelsets were built using geometrical scaling with the factor $\lambda = 20$. Such scale allowed studying the main physical phenomena taking place in the system under consideration [2].

To assess the characteristics of the modular wheelset in mock-up and field testing, second competitive car truck model with standard wheelsets was used [4]. This helps to distinguish real advantages of the modular design of the wheelset and provides adequate estimates of mock-up and field tests independently of numerous factors affecting the dynamic influence in the "wheelset – track" system [10]. The testing process using physical models has heuristic nature. In this connection (in accordance with the goal and plan of mock-up testing), a technique and system for documenting the measurement results as special tables was elaborated.

Figure 2. Models of track and car truck testing site.

The comparative testing of competitive modular wheelsets was performed varying the following parameters of the "wheelset – track" system [8]:

– number of waves of horizontal irregularities on one of the rails on a direct length of track (in real measurement, the horizontal irregularity wave length l = 1.1 m and amplitude h = 0.001 m);

– width of the track on a direct length of track;
– number of waves (n) of undulating wear on one of the rails on a direct length of track;
– nonparallelism of the car truck wheelsets during movement on a curved and direct section of rail.

The vibratory acceleration of the models of the car truck and rail elements, as well as the noise generated by their movement were measured by a VShV-003-M2 noise and vibration meter.

The trajectories of the car truck models were recorded using a laser source mounted on the truck chassis. The laser beam was projected on the screen where its trajectory was recorded. The movement dynamics of the truck models was compared using maximum amplitudes and frequencies of the oscillation spectrum of the "wheelset – track" system elements.

Figures 3a, 3b and 3c, correspondingly, depict the trajectories of the rail track profile and those of car trucks with conventional and modular design on the wheelsets. The trajectories were recorded using a laser source by two angular coordinates: pitching (ϕ_T) and hunting (ψ_T). According to Figures 4a – 4c, the trajectory of the car truck with conventional wheelsets in the horizontal plane along coordinate (ψ_T) has standard deviation in relation to the rail axis trajectory which equals $\delta_{\psi_T} = 0.86\,deg$. That of the truck with independently rotating elastic gaskets flanges $\delta_{\psi_T} = 0.487\,deg$. This denotes significantly lower level of the horizontal dynamics [5] and larger stability of the truck movement having the independently rotating elastic flanges [9].

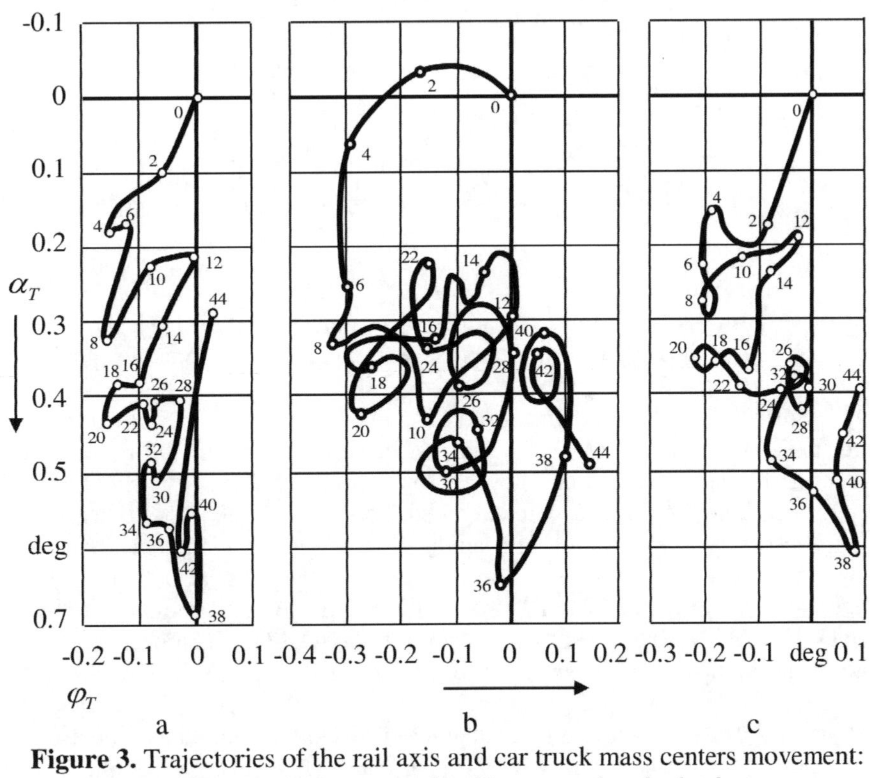

Figure 3. Trajectories of the rail axis and car truck mass centers movement:
a) rail track axis; b) car truck with conventional wheelsets;
c) car truck with modular wheelsets.

The models with modular wheelsets feature free oscillating boxes with three clear amplitude peaks on the acceleration spectrum at 28, 112 and 447 Hz, while the natural oscillations of the conventional wheelset have only one peak (28 Hz). During the movement under forced oscillations, the acceleration spectrum amplitude of the modular wheelset increases without changes in frequency. In the same conditions, the acceleration spectrum of the conventional wheelset changes its frequency content: in addition to the peak at 28 Hz, two more peaks occur at 112 and 447 Hz. Such difference between the spectra of natural and forced oscillations for the conventional wheelset is conditioned by the distribution properties of the track [3].

According to the plots (Figure 4, curves 1 and 2), the amplitudes of vibrational accelerations at high frequencies for the new wheelset design are 1.5 – 2 times lower than that of the conventional wheelset. This plot also presents the spectrum of the vibrational acceleration of the conventional wheelset box (Figure 4, curve 3) recorded in real operational conditions.

The results of the experiments on three variants of wheelset design have shown that all obtained spectra of vibrational acceleration (Figure 4, curves 1, 2 and 3) have characteristic coincidence of natural oscillation frequency zones of the studied models and real sample (Figure 4, curves 1, 2 and 3) that are in a range of 25 – 30 Hz recorded at the speed of 70 km/h. This indicates the satisfaction of the kinematic similarity criterion for the model and real "wheelset – track" systems [6].

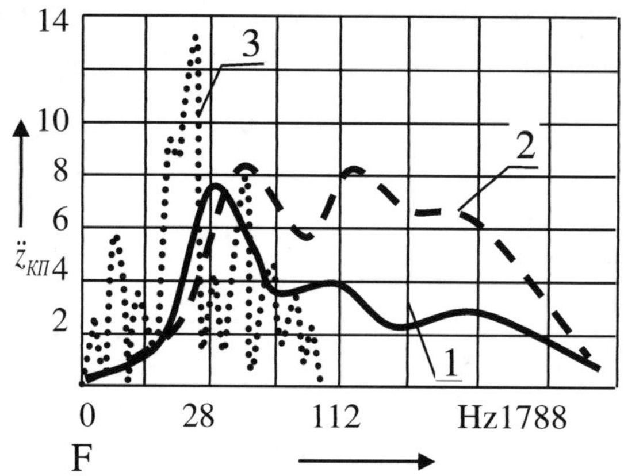

Figure 4. Spectrum of vibrational acceleration of the wheelsets:
1) new design; 3) conventional de-sign; 3) real conventional design.

Within the specified financial and time limits for manufacturing and field tests of the modular wheelset, three copies were made with rigid design of the carrying wheel – all-rolled wheel (Figure 5). In 2016-2017, at the Experimental course of Shcherbinka railway station, field tests of this variant of wheelset mounted on a cargo car truck were executed. Preliminary bench static tests were performed. The schemes of tensoresistor installation and application of testing loads are presented in Figure 5. Figure 6 shows the modular wheelset on a testing bench. The vertical load to the axle neck was applied in stages within a range of 30 – 200 kN, while the range of horizontal forces ap-plied to the wheel flange was 20 – 73 kN. The time of each loading stage was 3 – 5 seconds.

The measurement results have shown that there is no residual strain in the structure. The largest stresses in the most loaded zones of the wheels and axle ranged from 70 to 90 MPa, while the permissible value is 290 MPa. The strength characteristics of the modular wheelset correspond to effective regulations of the Russian Railways Company and allowed installing them in the cargo car truck to perform a set of tests [7].

Stationary braking tests were conducted for a semi-car, which first car truck had modular wheelsets installed, while the second car truck had conventional wheelsets. The effect of the pneumatic brake after complete service braking with consequent brake release was checked on an empty car and with simulation of its complete load. Following the results of stationary braking tests, the semi-car with the first car truck with modular wheelsets was allowed for running tests at the Experimental Course of the Railway Research Institute "VNIIZHT".

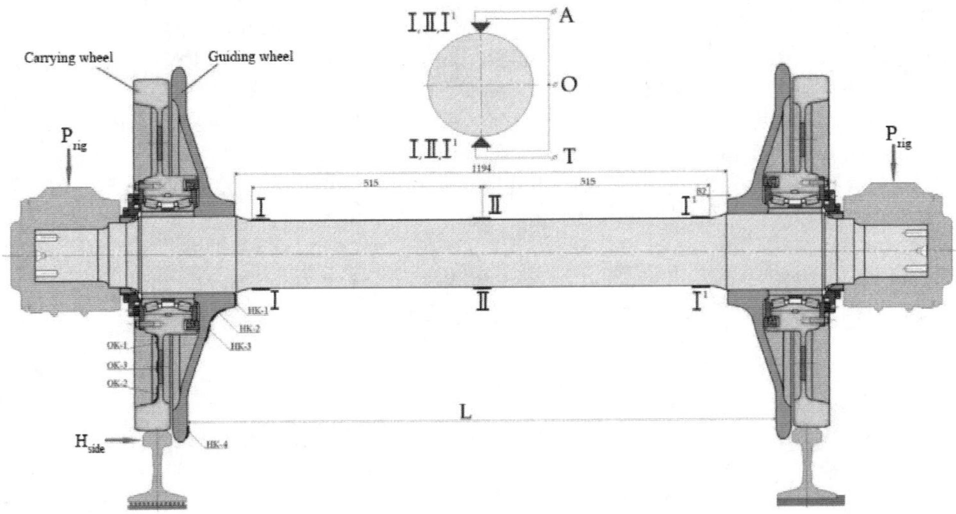

Figure 5. Scheme of testing load application and arrangement of tensoresistors in bench tests.

Figure 6. Loading bench tests of the modular wheelset.

On the next stage of field tests, the modular wheelset was tested on the track, which was destined to assess the functional capabilities: performance on a direct and curved sections of the track and safe railway operation. After the semi-car with the modular wheelset in the first car truck was tested, it was allowed for further testing – rides on circular tracks with speeds up to 120 km/h. The side frames of both car trucks were equipped with sensors to measure horizontal (frame) forces and coefficients of vertical dynamics [5]. The schemes to determine the coefficients of vertical dynamics were calibrated by lifting the body by jacks until complete unloading of the car trucks. The system was calibrated to determine frame forces by applying a specified horizontal load to the sides of the truck frames.

3. Results and discussion

The field test results of the modular wheelset on the Experimental Course of the Railway Research Institute "VNIIZHT" in Shcherbinka established that the dynamic indicators of the truck with the modular wheelset meet the regulations of the Russian Railways Company. The coefficients of vertical dynamics along the empty truck frame did not exceed 0.83 (with permissible value of 0.9). For the loaded truck, the coefficients did not exceed 0.39 (with permissible value of 0.85). The coefficients of horizontal dynamics amounted to 0.36 and 0.26, correspondingly for empty and loaded truck (permissible values are 0.4 and 0.38, respectively). The coefficients of stability margin were not less than 1.47 for empty and not less than 1.88 for loaded truck with the permissible value not less than 1.3. The vertical accelerations of the car center plate over the first truck were not more than 0.38g for empty car and not more than 0.29 for loaded one. Over the second truck (with conventional wheelsets), these values amounted, correspondingly, not more than 0.56 g and 0.42 g.

All dynamic indicators of the truck with the modular wheelsets received "good" score. For empty car, some of the dynamic indicators of the truck with conventional wheelsets were 30 – 40% worse and corresponded to "satisfactory" score. For loaded car, both trucks received "good" score, while the dynamic indicators of the truck with conventional wheelsets were 10 – 25% worse.

4. Conclusions

1. Based on the results of static, braking and running tests on the Experimental Course of the Railway Research Institute "VNIIZHT" it was established that the strength and braking indicators of the testing car trucks with modular wheelsets correspond to the requirements of the "Norms of calculation and design of railway cars of the Ministry of Railways with the rail gage of 1520 mm" and Guiding Document 24.050.37.

2. The modular wheelset with rigid carrying wheels with the second cargo car truck with convention-al wheelsets has 10–40% lower dynamic indicators as compared to the conventional wheelset.

Acknowledgments

This work has been supported by RVC Seed Fund, Ltd for Flexible Wheel, Ltd.

References

[1] New wheelset design for rail transport 2012*Bull. of the Institute of Natural Monopoly Problems: Railways Equipment* ISSN 1998-9318 Moscow **4 (20)** 64-72

[2] Galiev I, Shiler A, Kildibekov A and Kukushkin I 2015 25th DAAAM International Symposium on Intelligent Manufacturing and Automation (Vienna, Austria) **100** 1724-31

[3] Shiler A 2015 25th DAAAM International Symposium On Intelligent Manufacturing And Automation (Vienna, Austria) **100** 1714-23

[4] Casanueva C, Alonso A, Eziolaza I and Gimenez G 2012 Proceedings of the Institution of Mechanical Engineers *Journal of Rail and Rapid Transit*

[5] Stefano B and Gonzales B L Flexible wheelset-track model for numerical estimation of stresses in railway axles, Anno Accademico 2009-2010

[6] Petrov G and Tarmaev A 2018 Proceedings of the International Conference "A via mechanical Engineering and Transport" (AVENT) Advances in Engineering Research (Atlantis Press) vol **158** 410-415

[7] Borodin A and Prokofieva E 2017 Methods of substantiation of specialization of railway lines *Transport Problems* **12** 35-44

[8] Meymand S Z, Keylin A and Ahmadian A 2016 Vehicle System Dynamics (Moscow) **54(3)** 386-428

[9] Donzella G., Faccoli M., Mazzù A., Petrogalli C and Roberti R 2011 Progressive damage assessment in the near-surface layer of railway wheel – rail couple under cyclic contact *Wear* **271(1)** 408-416

[10] Gusak G V 2018 Reaction Forces under Static Loading of a Wheel Pair with Camber *Moscow University Mechanics Bull.* **73(4)** 84-90

[11] Eadie R and Old know D Rail grade selection and friction management: A combined approach for optimizing rail-wheel contact 2013 *Ironmaking and Steelmaking* **40(2)** 108-114

Studies of Current Modulation in the DC Voltage Link of the Frequency Converter during its Operation with Asynchronous Load

O V Arsentiev and G O Arsentiev
Irkutsk National Research Technical University, 83, Lermontov Street, Irkutsk, 664074, Russia

E-mail: arsentyevov@mail.ru

Abstract. The paper discusses current issues on the inverters peculiarities in the frequency converter. Theoretical studies have been carried out to determine the influence of the induction motor mode on the current modulation in the DC link. The authors suggest a calculation technique to determine the form of transient current characteristics when exposed to a single pulse of the inverter. Model studies of the system, calculation and construction of current characteristics proved the correspondence of theoretical and practical results. The research results determined the relationship between the current modulation in the DC link and the induction motor mode of operation.

1. Introduction

Mechatronic systems are the basis of modern production. They allow implementing a whole range of interrelated actions, movements and transitions united into a single technological process. The primary link of such a system is a controlled electric drive (ED), which moves the executive bodies of the working machines. At the same time, the electric drive configuration, the degree of semiconductor converting devices (SCD) application, and the degree of numerical control determine the possibility of regulating the process and controlling ED [1-3]. Thus, there is dependence between the technological process, the method of organizing a controlled electric power supply and the possibility of its integration into the general production automation system.

A common way to construct a controlled electric drive system is to implement the electric control method, when the magnitude and mechanical movements are carried out by regulating the parameters of the CD eclectic energy, with its subsequent supply to the electromechanical converter - electric motor (EM) [2-6]. The basis of modern controllers is a semiconductor converter, which must correspond to the external power supply system, an electric motor type, the possibilities to organize additional control by introducing feedback signals characterizing the output coordinates of the electric power and process parameters [6-9].

At present, the controlled electronic drive is mainly implemented in production by frequency controllers with asynchronous EM (FC-AD). Such a system is universal and allows organizing different ways of controlling separate, group and interconnected mechanical movements. The use of an asynchronous EM with a squirrel-cage rotor increases the reliability of the electromechanical system [8-12]. The study of the operation of the systems allows expanding their functionality; formulate recommendations for their use in electronic information technology.

IOP Conf. Series: Materials Science and Engineering **795** (2020) 012014 doi:10.1088/1757-899X/795/1/012014

2. Theoretical studies of current modulations in the FC-AD system

The basic schema of the frequency CD is a two-link system, with an autonomous voltage inverter. Studies have shown that there is an undoubted connection between the current modulation in the DC link with the operation mode of the asynchronous EM [11], the ratio of the active and reactive power of the consumer. To identify the basic laws of the process, it is necessary to consider the process of the inverter when implementing a three-phase voltage system.

Let us consider the process of a single pulse working out of the inverter key. Non-stationary phenomena in circuits with lumped parameters make the theoretical basis of the study. Figure 1 shows the design scheme that determines the inverter key influence on the active-inductive load (figure 1a) and the characteristics of the transient process (figure 1b), whose analysis allows determining the main patterns occurring in the DC link in the FC-AD system.

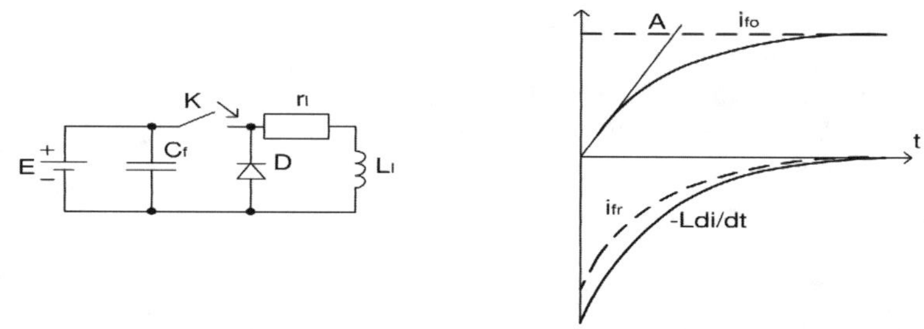

(a) (b)

Figure 1. Calculation scheme for the inverter key in the switching mode (a) and a general view of the transient characteristics (b).

The following elements and quantities are present in the circuit:

E – emf corresponding to the voltage in the DC link;

C_f is a capacitive filter to smoothen ripple in the DC voltage circuit;

K is a switch simulating the operation of an inverter key;

D is a reverse diode forming a circuit with the key turned off;

r_l, L_l are the resistances determining the asynchronous EM mode of operation

Characteristics determine the transient process when K is turned on. In general, the current flowing through a load simulating an EM equals:

$$i = i_{fo} + i_{fr}\,, \qquad (1)$$

where: i_{fo} is the forced current component corresponding to the steady state;

i_{fr} is the free component of the current, maximum at the moment of switching on K ($t=0$) and tends to 0 as $t\to\infty$, is directed opposite to the forced component of the current i_{fo};

$-Ld\frac{di}{dt}$ – emf of inductance, at the moment of switching on K is equal to the voltage in the DC link E, and tends to 0 as $t\to\infty$.

The current rise rate at the moment of switching on K ($t=0$) is determined by

$$\left(\frac{di}{dt}\right)_{t=0} = \frac{E}{L_l}\,.$$

Graphically, it is determined by the tangent tilt to the characteristic$i_{fr}f(t)$, its intersection with the line i_{fo} (point A). It is evident that the tangent tilt does not depend on the active load r_l.

Based on the equation (1), the current via the transient load can be determined by the expression

$$i = \frac{E}{r_l} - \frac{E}{r_l} e^{-\frac{r_l}{L_l}t} \qquad (2).$$

When the current changes, magnetic energy is accumulated in the reactance of the circuit L_l, i.e. reactive power consumption occurs according to the change in the transient component i_{fr}. Based on this, we can write an expression for determining the energy consumed by the inductance

$$A = \frac{L_l}{2} i_{fr}^2,$$

or with the account of (2)

$$A = \frac{L_l}{2}\left(-\frac{E}{r_l}e^{-\frac{r_l}{L_l}t}\right)^2.$$

As the current in the switching circuit changes, the potential energy of the magnetic field increases. It depends on the parameters of the circuit and the time of the transient.

Instantaneous AC Power in inductive coil is

$$q = ui_{fr}.$$

For the case under consideration, this equation can be written as

$$q = \frac{E^2}{r_l}e^{-\frac{r_l}{L_l}t}.$$

From this equation, we can conclude that the reactive power generation during the FC-AD system operation occurs as a result of the current modulation in the DC link. In this case, the nature of the inductance charge process corresponds to the curve of change $-Ld\frac{di}{dt}$ (figure 1b).

3. Experimental studies of current modulations in the FC-AD system

The empirical research was carried out with the numerical experiment method, applying the MatLab program. The FC-AD system was simulated with a block of parameters included for the study of current modulations in the DC link of an autonomous inverter [2, 11]. As a load, an asynchronous EM with a power of 125 kW and a synchronous rotation speed of 1500 rpm was used.

Figure 2 shows the operation of the inverter when generating the output voltage. The state diagram of the inverter switches (figure 1a) corresponds to the output three-phase voltage (figure 1b) obtained as a result of the simulation of the FC-AD system [11].

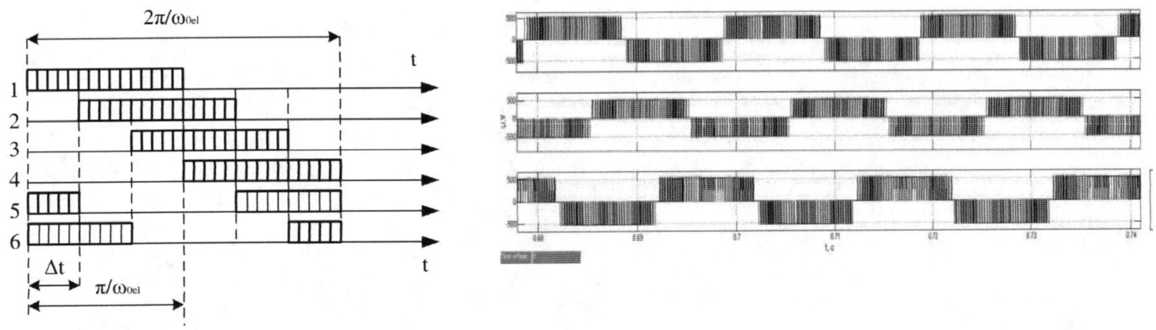

(a) (b)

Figure 2. The inverter works when generating the output voltage: a) a state diagram of the inverter keys, b) an output three-phase voltage system.

The main studies were carried out at a standard frequency of 50 Hz and a rated supply voltage of 0.4 kV. The oscillation period was $2\pi/\omega_{0el} = 0.02$ s, the time shift in the operation of the inverter keys, $\Delta t=0.0033$ s correspondingly. The above diagram and a graph of three-phase voltage correspond to the results received [1…6, 11] and are the basis for further research.

To determine the main dependences of the current modulation in the DC link on the load nature, model studies of the FC-AD system were carried out. Figure 3 shows the current characteristics for the load mode P = 81 kW, Q = 60 kW, which corresponds to $cos\varphi$ = 0.8, with a probable load of 80%.

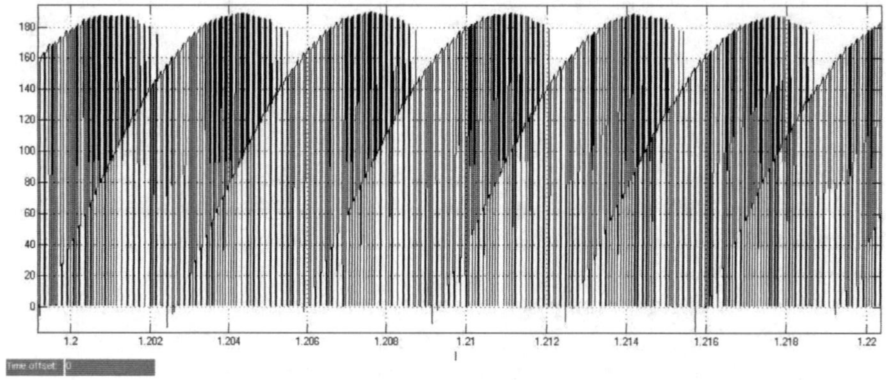

Figure 3. Current modulation in the DC link.

Analysis of current curves shows that the time shift between them is 0.0033 s, which fully corresponds to the time diagram of the inverter keys (Δt value, figure 2a).

The steady-state current value (maximum value) is 185 A and is confirmed by the calculation data received for this mode according to the above procedure. This value, with an error of 10%, corresponds to the forced component of the current $\boldsymbol{i_{fo}}$ for the considered mode, 169 A.

According to (2), the calculated equation is

$$i = 169 - 169e^{-479t},$$

where E = 400 V, r_l = 2.37 Ohm, L_l =4.6·10^{-3} GN.

Figure 4 shows the results of the calculation for the current modulation curve $i_{fr}f(t)$ in the DC link for the considered mode of the FC-AD system.

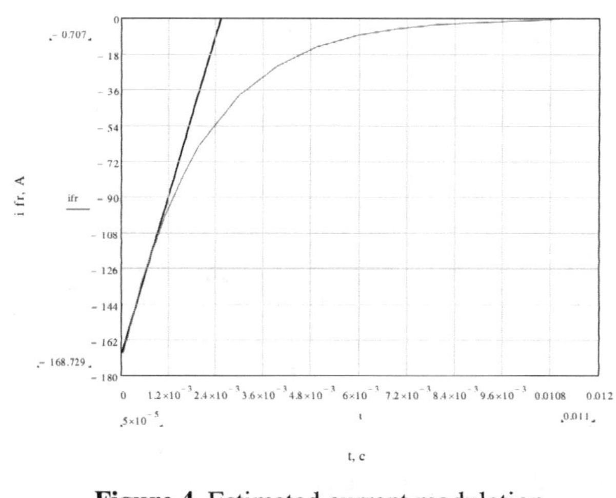

Figure 4. Estimated current modulation.

The obtained characteristic shows that the current change period, when the inverter key is turned on is 0.0108 s, which is comparable with the pulse duration of the public key 0.01 s. The tangent to the initial phase of the characteristic shows the minimum time i_{fr} changes to reach the steady-state value

i_{fo}. The physical meaning of this process is theoretically determined as ((·) A, figure 1b), $t_{min} = 2.6 \cdot 10^{-3}$ s. The characteristic of the free current component in the DC link is dying out, varies within -169 $> i_{fr} > 0$. In general, the calculated current change curve reflects the physical processes in the DC link of the inverter.

Its differences from the experimentally determined current modulation (figure 3) are due to three simultaneous pulses. They are shown by the diagram of the inverter keys. This overlapping distorts the final pattern of the pulses distribution and does not allow determining the effect of a single key switch.

4. Conclusion

Theoretical studies have made it possible to predict the current modulation when powered by a constant voltage. It has been determined that the parameters of the switching circuit - the active and reactive resistances of the asynchronous load, the voltage value in the DC link have a key influence on the current transient process. In this case, the current modulations have a forced i_{fo} and a free component i_{fr}. The nature of the change in reactive power at the inductive-active load of the inverter has been determined. The model studies confirmed the main theoretical results with an error of no more than 10%.

References

[1] Wang J, Huang A Q, Sung W, Liu Y and Baliga B J 2009 *IEEE Industrial Electronics Magazine* **3(2)** 16–23

[2] Tety P, Asseu O, Diby A, Konan H and Gbegbe R 2017 *East Journal of Electronics and Communications* **17-5** 1-12

[3] Bose B K 2000 *IEEE Trans. Power Electron.* **15-4** 688-701

[4] Blaabjerg F, Chen Z and Kjaer S B 2004 *IEEE Trans. Power Electron.* **19-5** 1184-1194

[5] Giridharan K, Chellamuthu C, Kannabhiran A. and Santhi R 2016 *East Journal of Electronics and Communications* **16-4** 859-879

[6] Balaji R, Karthick V, Thulasi J A 2015 *Journal of Advanced Engineering* **2-2** 46-50

[7] Walker G R and Serbia P C 2004 *IEEE Trans. Power Electron.* **19-4** 1130-1139

[8] Arsentiev O V, Baranov P R, Vilnin A D, Kladiev S N 2017 *Bulletin of the Tomsk Polytechnic University, Geo Assets Engineering* **328 (11)** pp 126-138

[9] Guo X, Cavalcanti M C, Farias A M and Guerrero J M 2013 *IEEE Trans. Power Electron.* **28-6** 2635-2637

[10] Irwanto M, Alam H, Mashor M Y, Masri M, Haziah A H, Ismail B and Butar-Butar A H 2017 *Far East J. of Electronics and Communications* **17-5** 1167-76

[11] Arsentiev O V, Arsentiev G O, Dunaev M P 2018 *IOP Conf. Series: Materials Science and Engineering* **327** (2018) *052005*

[12] Xiao H and Xie S 2012 *IEEE Trans. Power Electron.* **27-4** 1799-1808

Mathematical Modeling of the Electric Drive with Universal Power Semiconductor Converter

M P Dunaev[1] and L V Arshinskiy[2]

[1]Irkutsk National Research Technical University, 83, Lermontov Street, Ikutsk, 664074, Russia
[2]Irkutsk State Transport University, 15, Chernishevsky Street, Irkutsk, Russia

E-mail: mdunaev10@mail.ru

Abstract. In the article are examined the mathematical models of the electric drive of direct and alternating current with the universal power semiconductor converter. The functional diagram of electrical power semiconductor converter is described. Are given the oscillograms of the output voltage of power semiconductor converter, obtained with the aid of the model of converter. Are determined the static and dynamic characteristics of the electric drive of direct and alternating current with the universal power semiconductor converter, which make it possible to positively estimate fitness for work and accuracy of the systems of electric drive in the steady-state and dynamic behaviors of work. The experimental studies, carried out in the laboratory mock-up of the electric drive of direct and alternating current with the universal power semiconductor converter, testify about the sufficiently high adequacy of the developed models and confirm the practical possibility of designing of the class of universal semiconductor converters for regulating the angular velocity of the electric motors of direct and alternating current.

1. Introduction

Power semiconductor converters (PSC) are one of most important components of the contemporary adjustable electric drive [1].

Power semiconductor converters are divided into two large groups: direct-current converters and the converters of alternating current. The first of the enumerated groups of converters is intended for regulating of direct current and voltage in the load, the second - for regulating of alternating current and voltage [2]. The basic types of direct-current converters it is possible to consider the controlled rectifier (CR) and pulse-width converter (PWC), whereas to the basic types of the converters of alternating current should be carried the regulator of voltage (RV) and frequency converter (FC).

The practice of the use of power semiconductor converters shows that in a number of cases it is convenient to have the certain universal power semiconductor converter (UPSC), which combines in itself the possibilities of regulating both a constant, and alternating voltage in the load. The development of this UPSC will make it possible to reduce the nomenclature of necessary PSC and to reduce the necessary quantity of the technical documentation.

Entire above-mentioned explains significant interest in UPSC, research on the schemes of which is convenient to carry out with the use of mathematical simulation.

IOP Conf. Series: Materials Science and Engineering **795** (2020) 012015 doi:10.1088/1757-899X/795/1/012015

The packet is the well recommended itself means of mathematical simulation MatLab [3, 4] with the widely developed applications (Toolboxes), from which Toolbox Simulink it is well fitted out for synthesis and analysis of different electrical systems [5].

Purpose of this work - a study with the aid of the mathematical models of the static and dynamic characteristics of the schematic of electric drive with UPSC proposed.

2. Mathematical models

For developing the mathematical model of electric drive with UPSC we will use the approach, presented in [6 -10]. To figure 1, where is shown the functional diagram of electric drive with UPSC, are designated the following blocks: V1, V2 - diode rectifiers, assembled according to three-phase bridge diagram; F – ripple filter; AIV – the transistor autonomous inverter of voltage; CSI – control system of inverter; M1 – alternating-current motor (induction motor with the short circuit by rotor); M2 – direct-current motor (with the separate excitation).

The diode rectifier V1 converts the alternating voltage of power line (~Up) into rectified constant voltage (=U), which then by ripple filter F (usually inductive-capacitance) it will be given in the form the voltage of supply (=Ud) on the inverter AIV. To the output AIV can be connected either the alternating-current motor M1 or through the rectifier V2 - the direct-current motor M2. Output frequency AIV (fout1) is regulated with the aid of microprocessor CSI, moreover the amplitude of output alternating voltage AIV (~Uout1) and the amplitude of the output constant voltage V2 (=Uout2) also is regulated in the block CSI with the aid of pulse-width modulation (PWC). As a rule, three-phase bridge diagram on the IGBT- transistors is used as the structure diagram of inverter.

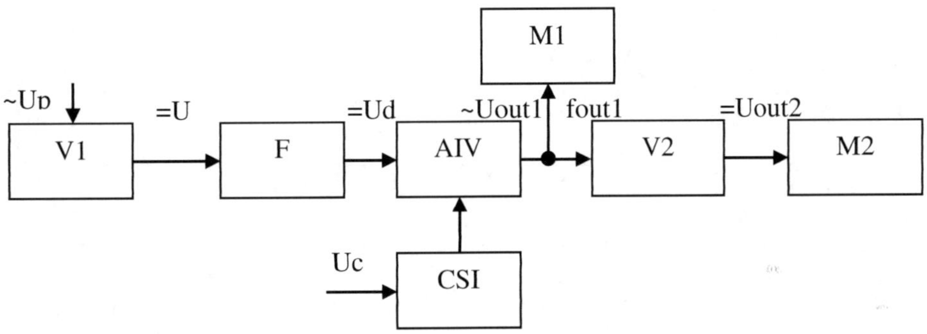

Figure 1. Functional diagram of electric drive with UPSC.

The virtual laboratory model of the system of electric drive “universal power semiconductor converter – induction motor” (further UPSC - AM) is given to figure 2 and is intended for studying the static and dynamic characteristics of this system.

The following blocks enter into the composition of model: dc power supply (DC Voltage Source); power part UPSC (Invertor IGBT); the system of administration UPSC (Invertor Control System); induction motor with the short-circuited rotor (Asynchronous Machine); the load of induction motor with the short-circuited rotor (Mechanical Load); the set of instruments for measuring of angular velocity, moment (Scope 1) and stress on the stator of induction motor with the short-circuited rotor (Scope 2).

Regulating the value of stress on the stator and frequencies of the rotation of engine is produced by changing the parameters CSI (Invertor Control System, the parameters Frequency of output voltage and Modulation index). A change in the moment of the load of the engine M1 (Asynchronous Machine) is produced by timing and value of the jump of the load (Mechanical Load).

IOP Conf. Series: Materials Science and Engineering **795** (2020) 012015 doi:10.1088/1757-899X/795/1/012015

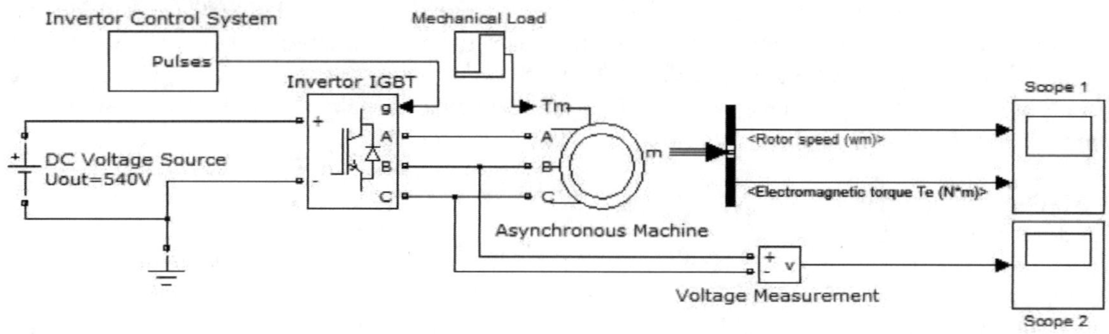

Figure 2. Model of the system UPSC-AM.

With the aid of the model, depicted to figure 2, was determined the oscillogram of voltage on the output of virtual model UPSC (figure 3). This oscillogram illustrates the method of pulse-width modulation for obtaining necessary of form and amplitude of the output voltage of inverter used.

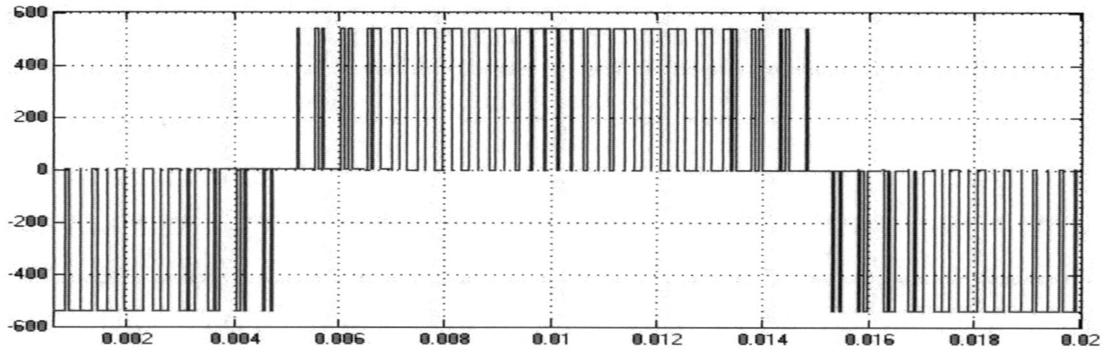

Figure 3. Oscillogram of the output voltage of model UPSC.

Also with the aid of the model, depicted to figure 2, was determined the regulation characteristic of output voltage UPSC on the idling Uout = f (fout), where fout - output frequency UPSC; Uout - voltage on output UPSC (on the engine). The obtained characteristic Uout = f (fout) is shown to figure 4.

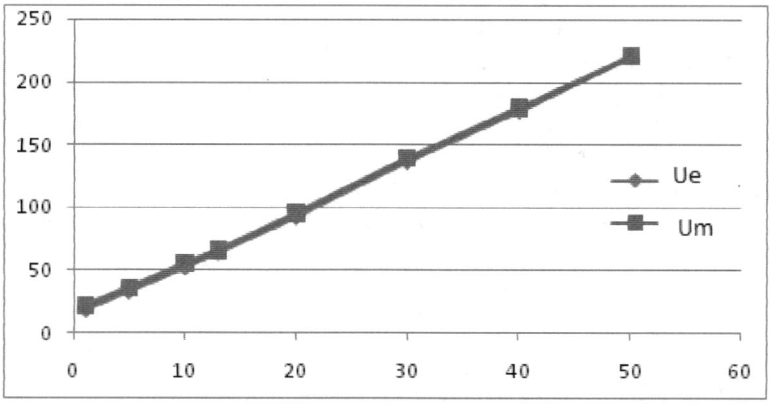

Figure 4. Regulation characteristic UPSC.

In addition to this with the aid of the model of figure 2 were determined the mechanical characteristics of the system UPSC-AM W_M= f (M) with the the fout =25 and 50 Hz. The data of characteristic are cited to figure 5 and are designated by index "m". The mechanical characteristics of system UPSC-AM go they in parallel and possess significant hardness.

The experimental studies, carried out in the laboratory mock-up of electric drive with UPSC, make it possible to speak about the sufficiently high adequacy of the model of figure 2. About this testify the experimental mechanical characteristics of system UPSC-AM, also shown to figures 4 – 5 and having the index «e».

As can be seen from the comparisons of model and experimental data of figures 4 and 5, an error in the mechanical characteristics engine, obtained with the aid of the mathematical simulation, does not exceed 5 %.

Thus, the constructed on program MATLAB mathematical models of system UPSC-AM with the adequate accuracy describe static behavior.

The dynamic engine characteristics in the system UPSC-AM W_M=f (t), M_M=f (t) and U_M =f(t) are shown to figure 6. Data of the characteristic of the engine W_M =f (t) in the system UPSC-AM testify about the stability of system UPSC-AM in the starting regime and the regime of the spasmodic rise of nominal load (M = 1,15 nm).

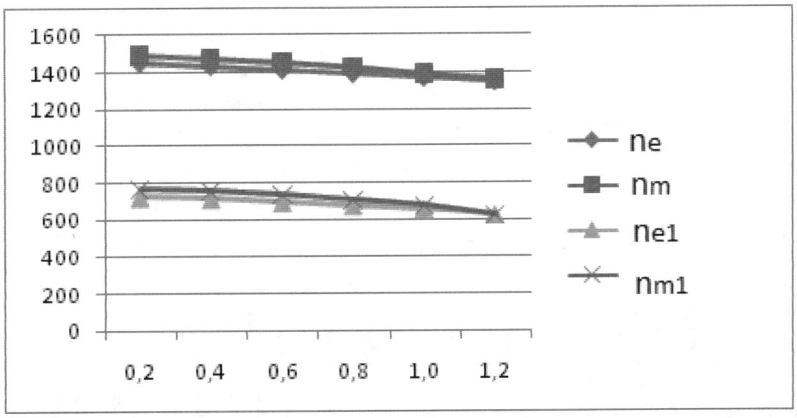

Figure 5. Mechanical characteristics of the model of system UPSC-AM: upper curves – with 50 Hz, lower curves – with 25 Hz.

Parameters of the transient process W_M =f (t) with an abrupt change in the controlling influence of fout s 0 to 50 Hz following: the starting time of t_p = 0,25 s, the overregulation σ_m = 0 %, static error with the rise of load (M = 1,15 nm) δ=9,8 %.

The virtual laboratory model of the system of electric drive "universal power semiconductor converter is – direct-current motor" (further UPSC-DCM) is given to figure 7 and is intended for studying the static and dynamic characteristics of this system.

The diagrams of voltages and output current of model UPSC-DCM are given in figure 8.

With the aid of the model of figure 7 were determined regulation Uout=f (fout) and electromechanical characteristics of the system UPSC-DCM Of W_M= f (I) with the the fout =25 and 50 Hz. The data of characteristic are cited respectively to figures 9 - 10 and are designated by index "m". The electromechanical characteristics of system UPSC-DCM go they in parallel and possess significant hardness.

The launch procedure of the model of direct-current motor in the system UPSC-DCM is demonstrated on the diagrams, given to figure 11. The diagrams of signals at the output of model UPSC-DCM (figure 8), static characteristics (figures 9 - 10) and the diagram of launching direct-current motor (figure 11) testify about the fitness for work of model UPSC-DCM, given to figure 7.

Parameters of the transient process W_M =f (t) with an abrupt change in the controlling influence of fout s 0 to 50 Hz following: the starting time of t_p = 3 s, the overregulation σ_m = 0 %.

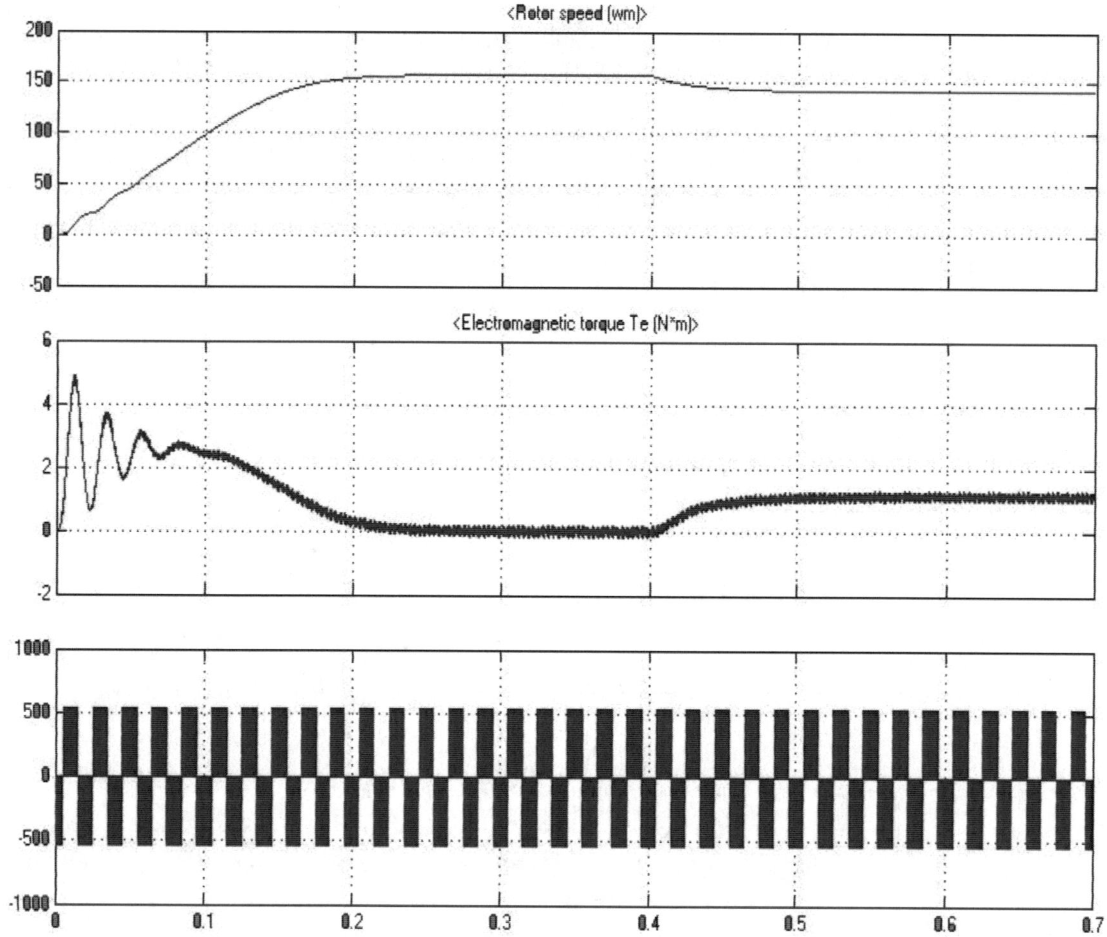

Figure 6. Dynamic engine characteristics in the system UPSC-AM.

IOP Conf. Series: Materials Science and Engineering **795** (2020) 012015 doi:10.1088/1757-899X/795/1/012015

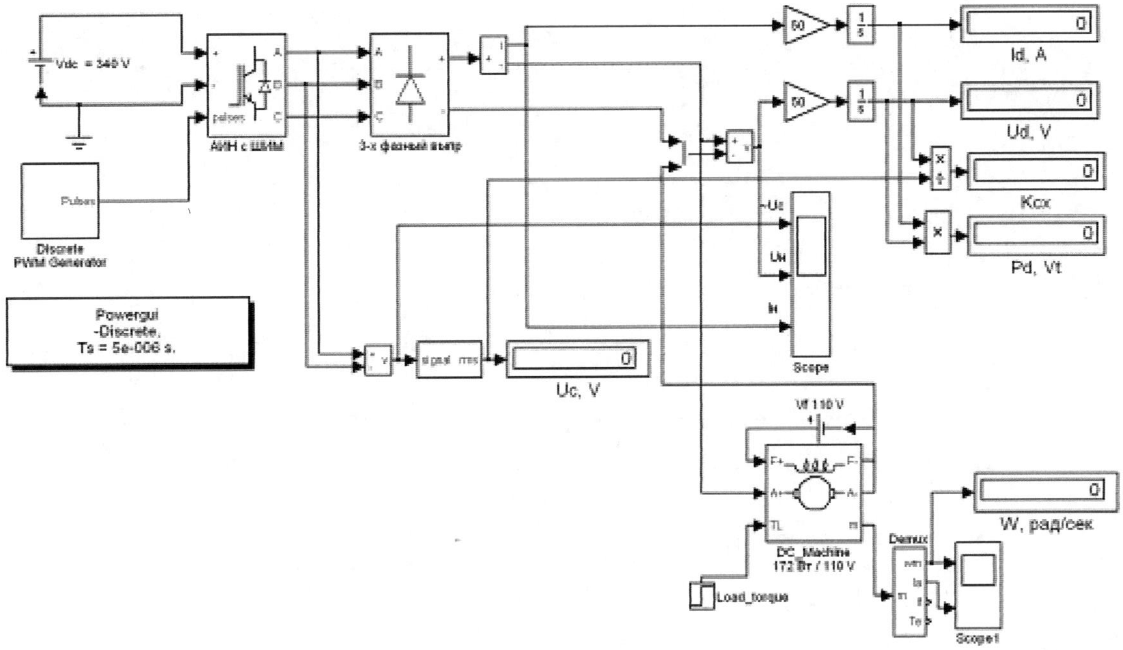

Figure 7. Model of the system UPSC-DCM.

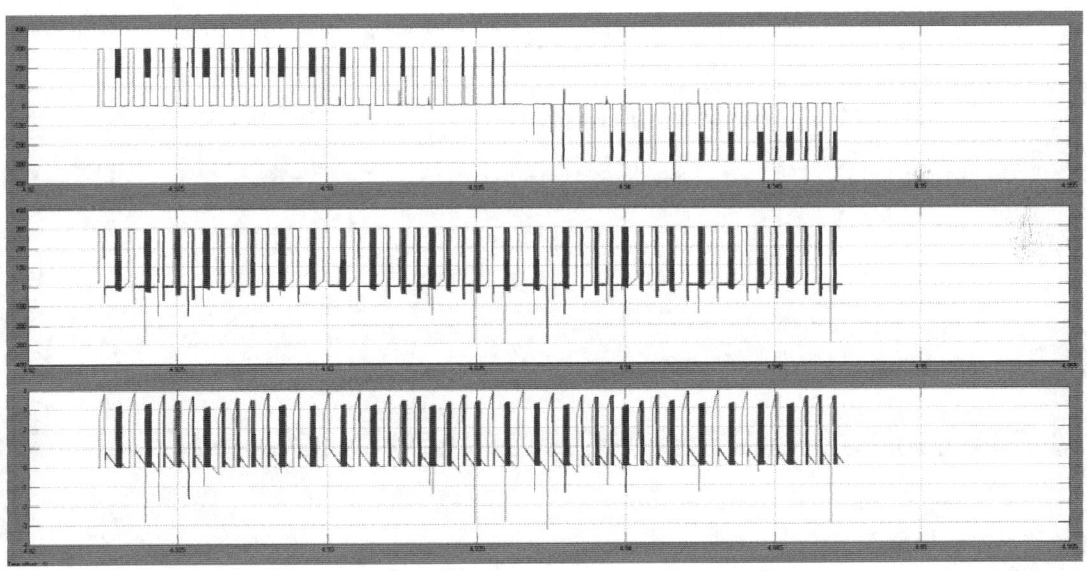

Figure 8. Diagrams of signals at the output of the model of system UPSC-DCM: upper is – voltage on output AIV, average - voltage on the output V2, lower – output current V2.

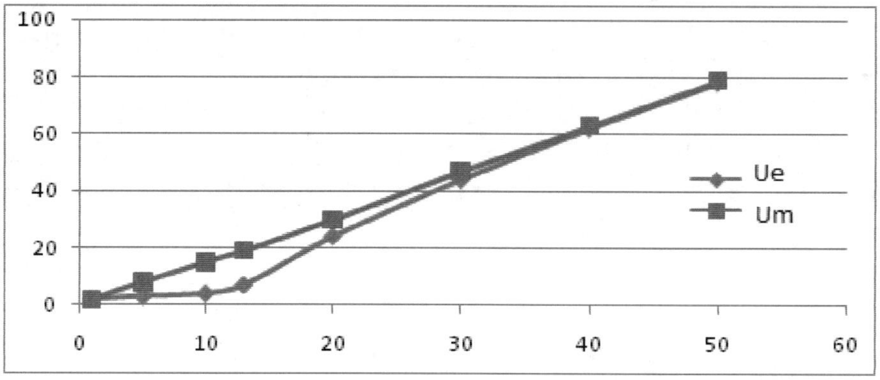

Figure 9. Regulation characteristic of system UPSC-DCM.

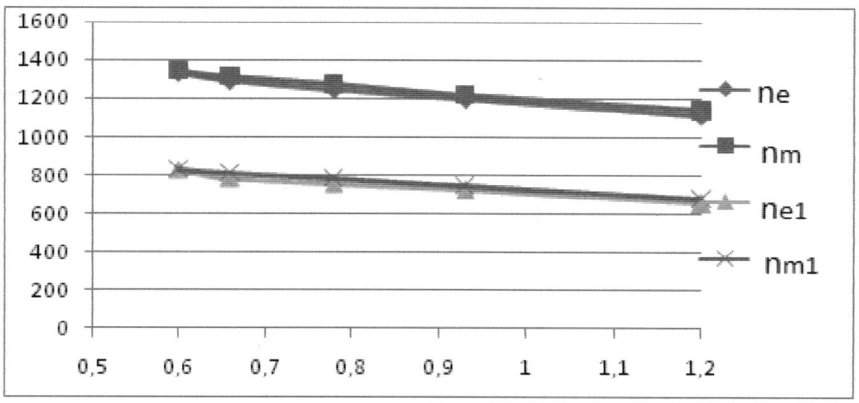

Figure 10. Electromechanical system characteristic UPSC-DCM.

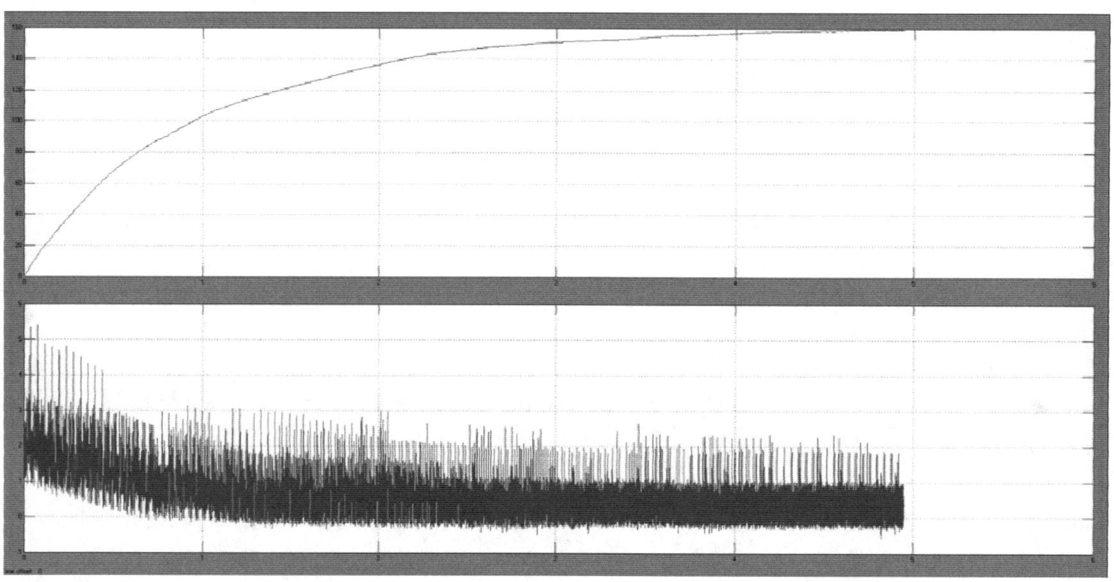

Figure 11. Diagrams of launching the electric motor of direct current in the system UPSC-DCM: the upper – angular velocity, lower is – the current of engine.

3. Conclusion

The experimental studies, carried out in the laboratory mock-up of electric drive with UPSC, make it possible to speak about the sufficiently high adequacy of the model of figure 7. About this testify experimental regulation and the electromechanical characteristics of system UPSC-DCM, shown to figures 8 – 9 and having index "e".

As can be seen from the comparison of model and experimental data figures 8 and 9, an error in the electromechanical engine characteristics, obtained with the aid of the mathematical simulation in the regimes, close to the nominal, does not exceed 5 %. Thus, the constructed on program MATLAB mathematical models of system UPSC-DCM with the adequate accuracy describe static behavior.

In the course of studies the static and dynamic characteristics of the systems of electric drive with the universal power semiconductor converter were determined. The represented characteristics make it possible to positively estimate the fitness for work of the systems of electric drive UPSC-AM and UPSC-DCM in the steady-state and dynamic behaviors of work.

References

[1] Bose B K 2000 *IEEE Trans. Power Electron.* **15-4** 688-701
[2] Blaabjerg F, Chen Z and Kjaer S B 2004 *IEEE Trans. Power Electron.* **19-5** 1184-1194
[3] Giridharan K, Chellamuthu C, Kannabhiran A and Santhi R 2016 *East Journal of Electronics and Communications* **16-4** 859-879
[4] Tety P, Asseu O, Diby A, Konan H and Gbegbe R 2017 *East Journal of Electronics and Communications* **17-5** 1-12
[5] Zakaryukin V P, Kryukov A V 2015 *Power Technology and Engineering* **49-4** 304-309
[6] Irwanto M, Alam H, Mashor M Y, Masri M, Haziah A H, Ismail B and Butar-Butar A H 2017 *Far East Journal of Electronics and Communications* **17-5** 1167-1176
[7] Balaji R, Karthick V and Thulasi J A 2015 *Journal of Advanced Engineering* **2-2** 46-50
[8] Walker G R and Serbia P C 2004 *IEEE Trans. Power Electron.* **19-4** 1130-1139
[9] Guo X, Cavalcanti M C, Farias A M and Guerrero J M 2013 *IEEE Trans. Power Electron.* **28-6** 2635-2637
[10] Xiao H and Xie S 2012 *IEEE Trans. Power Electron.* **27-4** 1799-1808

Organization of Quality Control of Manufactured Products in Smart Factories of the Industry 4.0

D A Zakoldaev[1], A V Shukalov[1], I O Zharinov[1], O O Zharinov[2]

[1]Faculty of Information Security and Computer Technologies, Saint Petersburg National Research University of Information Technologies, Mechanics and Optics, 49, Kronverksky Av., Saint Petersburg, 197101, Russia
[2]Department of Problem-Oriented Computing Complexes, Saint Petersburg State University of Aerospace Instrumentation, 67, Bolshaya Morskaia str., Saint Petersburg, 190000, Russia

E-mail: mpbva@mail.ru

Abstract. The task is to design instrument means of quality control of the items being manufactured in the Industry 4.0 item designing company. Item designing components manufacturing in the Industry 4.0 company is done automatically and that why the traditional schemes of quality check being used now in the Industry 3.0 companies cannot be used to organize new productions. There is a review of technical solutions which are used now to verify item designing components quality. There is a scheme and components which are necessary to check item designing components in the Industry 3.0 company. The advantages and disadvantages of existing procedures and equipment are described. There is a scheme and components which are necessary to check item designing components in the Industry 4.0 company where the cyber and physical systems and perspective cloud technologies, industrial Internet of Things technology are used.

1. Introduction

Automatized work places (AWP) with the nomenclature of the item designing components being checked can be classified into:

- single (narrowly specialized);
- group;
- universal.

Single AWP is a check scheme equipped technically and with program, mathematically, informatively, linguistically, organized and methodically to check, set and adjust a single unit of item being manufactured [1, 2]. All single type items are checked with single AWP in a sequence mode (in a row). The item designing components check being manufactured in the Industry 3.0 company is done automatizedly (with humans intervention) [3, 4].

Group AWP is a check scheme equipped with all necessary equipment to check simultaneously some single-type examples of the items being manufactured with a single control of checking connected to the instrument PC. If they apply group schemes of checking in an AWP it may reduce significantly the efforts of works being completed in the Industry 3.0 company and increase the volumes of items being produced keeping at the same time the item quality [5, 6].

A universal AWP is [7, 8] a scheme of checking equipped with all types of equipment to check simultaneously some items of different production nomenclature. The universal AWP key component is a control of checking and connection cables which may with minimal commutation of connection to link an instrument PC to check different types of items. In this case AWP software grants the operator to control the modes of item checking which are narrow and special in comparison to the group of items being checked. If we apply a universal AWP to check the items we may reach the results ten times faster in comparison to a single AWP.

To organize production division of new type [9] (Industry 4.0 automatic production) the new components and technologies [10, 11] must be designed and implemented which will be the base of new AWP equipped with a cyber and physical system to ensure the high quality of the items being produced.

2. Organization of item checking in the Industry 3.0 company

To check the items being produced an AWP has a set of control and measurement equipment which includes:

- amperemeter,
- voltmeter,
- oscilloscope,
- colourmeter,

and other technological equipment which helps the Industry 3.0 company personnel to detect the item defects, to provide the necessary technical specifications of items and other.

The designer of project company chooses among the units of control and measurement equipment or it also may be done by item manufacturer judging by metrology specifications of the equipment and characteristics of item being checked parameters. A list of recommended technological equipment to check the items from the designer as well as recommended schemes of checking are included into sets of construction (CD), program (PD) and technological (TD) documentations which later will be used to manufacture the item.

All types of measurement and control equipment which are parts of an AWP can be used as check schemes after a successful verifying. Control and measurement equipment have some check procedures which proves that the equipment technical specifications are according to the requirements which the equipment manufacturer stated in its technical passport and that kind of checking is being done once in two or three years. Checking controls and connection cables are being verified with marks in the technical passport as non-standard measures of measurement which requirements are given to the designers who design the items being produced.

Scheme and components to check item designing components in the Industry 3.0 company are given in figure 1.

When works of checking, setting and adjustments of manufactured items are completed in the Industry 3.0 company they initiate the item tests on the special technological equipment. Special technological equipment for test conducting includes the following types of testing chambers:

- chamber of heat and cold which checks the manufactured item resistance to the influence of high and low environmental temperatures;
- humidity chamber which checks the manufactured item resistance to the influence of high and low levels of air humidity (to create the necessary climatic conditions inside the humidity chamber it simultaneously changes the air temperature and humidity);
- pressure chamber which checks the manufactured item resistance to the influence of high and low levels of atmospheric pressure;
- atmospheric condensed rains chamber which checks the manufactured item resistance to the influence of different types of rains which appear in the environment including work solutions and aggressive media and other types of special chambers.

Also with testing chambers manufactured items quality control technological equipment includes a set of vibration stands which test the item mechanically. The mechanical tests prove that the item is resisted to different types of strike loads which influence the item in exploitation including mechanical strikes during the item transporting from the manufacturer plant to the exploitation point in all types of transport (maritime, automobile, railroad, aviation transport).

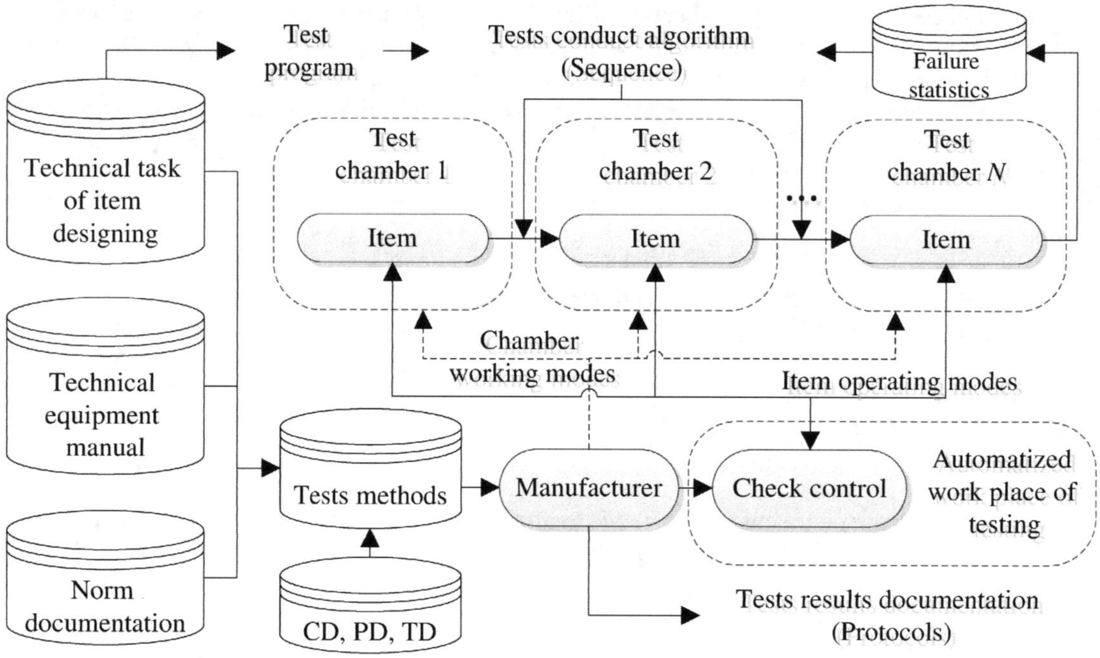

Figure 1. Schemes and components to check item designing components in the Industry 3.0 company.

3. Organization of item checking in the Industry 4.0 company

Control and measurement equipment in the Industry 4.0 company is known as cyber and physical systems (CPS) as items of optical and X-ray control, functional control (flying tests) and other to check automatically the manufactured items in all stages of technological operations. Item transporting to the work place of controlling CPS is done with robotized transport system which moves items within technological section (workshop) of the company from one working CPS to another.

Scheme and components to check item designing components in the Industry 4.0 company are given in figure 2.

The Industry 4.0 production company technological line is formed with a family of CPSs which connection topology is organized according to the CPS self-organization algorithms. In figure 2 there is an option how to unite CPSs with scheme «point-point» which forms the hybrid production line functioning automatically. Production machines of this line have two types of CPSs:

- CPSs to complete technological operations of item manufacturing (SMD (Surface Mount Device)-components installation on a printed circuit board, 3D-printing parts and other);
- CPSs to complete operations of production control which compare the manufactured item with the requirements of technical documentation.

CPSs of two types in linear schemes of connections of production machines is done in sequence to conduct the operation-wise control of the items being manufactured. CPSs to complete technological operations may independently as well to conduct technological procedures done quality control inside the work chamber with integrated system of automatic control. So operation-wise control of the items being manufactured in the Industry 4.0 company include the automatic procedures of:

- integrated control done with technical means of CPS (for example, quality control of separate layer application in 3D-printing on a substrate);
- operation control done with other CPS to check item in general for accordance to the technical documentation requirements.

IOP Conf. Series: Materials Science and Engineering **795** (2020) 012016 doi:10.1088/1757-899X/795/1/012016

After completion of all forms of production control they make an electronic tests protocol which is documented in the company cloud storage.

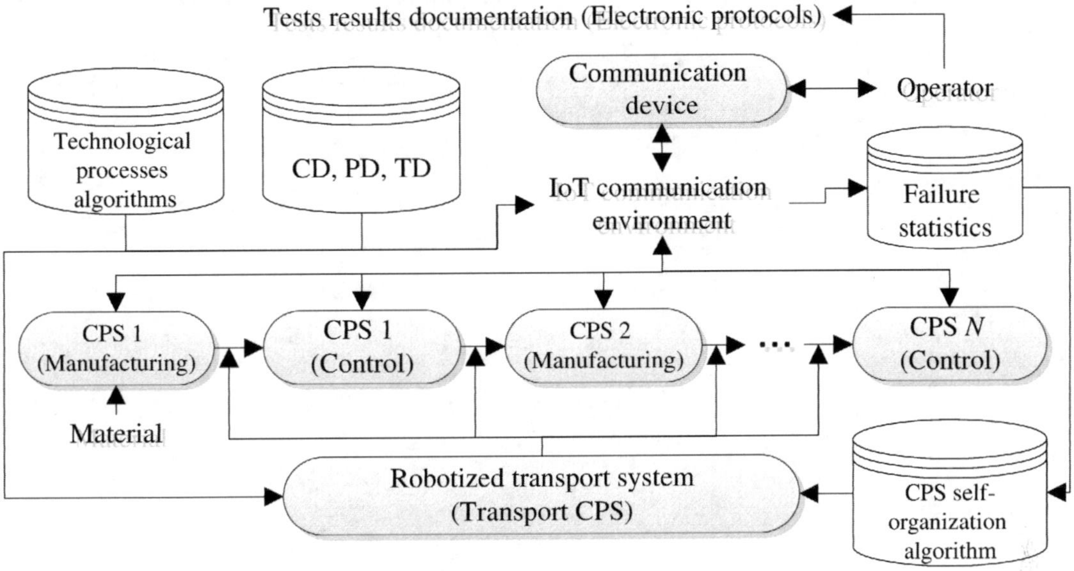

Figure 2. Schemes and components to check item designing components in the Industry 4.0 company (IoT - Internet of Things).

4. Conclusion

The designer does not define the order of test conduction in technical documentation for the items because in practice the manufacturer decides on their own the item tests order. So the manufacturer must choose the order of item checking which grants the maximum loads of the items being checked mode in different criteria to guarantee the consumer the item high quality speaking about the item which will be exploited.

Item load coefficient for each type of tests can be defined by the manufacturer being based on statistics data of item failures when all types of tests are conducted for a particular period of time. So item load coefficient during the tests will show the probability of failure which can be used to make an algorithm of item checking in an item test scenario with more load.

Mathematically the item check algorithm is a theory of graphs where the graph tops are equal to each type of tests and oriented ribs of graph is test sequence which numerically define the probability of item failure and time for each test conduct. Initial data for mathematical means of CPS self-organization algorithms is statistics data of item failure for each type of production tests to form the failure statistics in the company.

References

[1] Bonilla S H, Silva H R, Terra da Silva M, Goncalves R F and Sacomano J B 2018 *Sustainability* **10**(10) 3740
[2] Mabkhot M M, Al-Ahmari A M, Salah B and Alkhalefah H 2018 *Machines* **6**(2) 23
[3] Jerman A, Bach M P, Bertoncelj A 2018 *Machines* **6**(3) 41
[4] Baum J, Laroque C, Oeser B, Skoogh A and Subramaniyan M 2018 *Machines* **6**(4) 54
[5] Zeid A, Sundaram S, Moghaddam M, Kamarthi S and Marion T 2019 *Machines* **7**(2) 21
[6] Gurjanov A V, Zakoldaev D A, Shukalov A V and Zharinov I O 2018 *IOP Conference Series: Materials Science and Engineering* **327** 022110
[7] Zakoldaev D A, Shukalov A V, Zharinov I O and Zharinov O O 2018 *Journal of Physics: Conference Series* **1015** 052033

[8] Serrano W 2018 *Smart cities* **1**(1) 134-154
[9] Birkel H S, Veile J W, Müller J M, Hartmann E and Voigt K-I 2019 *Sustainability* **11**(2) 384
[10] Savastano M, Amendola C, Bellini F and D'Ascenzo F 2019 *Sustainability* **11**(3) 891
[11] Schallmo D, Williams C A, Boardman L 2017 *International journal of innovation management* **21**(8) 1740014

Method of Calculating of Spiroid Gear Resource by Wear at Step Loading Mode

A V Zaitsev

Siberian Transport University 630049, Russia, Novosibirsk, st. D. Kovalchuk, 191

zaitsev.zaw@yandex.ru

Abstract. The brief analysis of reasons of failure of mechanisms and drives of lifting, construction, and road-building machines based on engagement drives, and worm-gear class drives, represented by contact destructions of active surfaces of cog-wheel cogs resulting in malfunctions, breakages, failures, such as wear and furrows, was carried out. The need to create a method for calculating the wear of spiroids with regard of variable loading mode and time was substantiated. The method developed allows, with regard of real modes of operation of handling machinery, equipment and machines, determine the wear intensity values and calculate the spiroid gear drive resource with regard of duration of action of diving torque values in accordance with the set variable loading schedule for lifting, construction, and road-building machines under the wear intensity of the spiroid wheel cog on values of driving torque on the spiroid gear drive output shaft, obtained from experiments.

1. Introduction

Due to the action of useful resistances, constant or variable in time in the process of operation, occurring on operating equipment, contact destructions of active surfaces of links, wear, furrows on cog side surfaces and fractures, which are the reason of failure of mechanisms of drives of lifting, construction, and road-building machines, form in cog engagements of power transmissions [1]. This type of destruction is typical for quenched surfaces of cog-wheel cogs and is revealed by progressive pitting of material that may reach considerable size, and on work-hardened cog surfaces by flaking lage areas of metal pad [2].

For cog wheels of both open and closed drives, and worm-gear class drives operating in the environment contaminated with abrasive particles, the typical reason of failure is wear and binding of cogs.

The typical feature of worm-gear class drives is the relative slippage at any point of contact of active surfaces of the couple links, the speed value of which increases with the transmission ratio increase.

Mechanisms of drives of lifting, construction, and road-building machines are characterized by intermittent-duty loading modes with the change of drive torques on the gear drive output shaft during the operating cycle time. There modes are characterized by low line speed and related high loads, along with frequent startups and high contacts voltages leading to destruction of the surface of cogs of contacting links and, consequently, wear.

Spiroids belonging to worm-gear type (see figure 1 [3, 4, 5, 6, 7, 8]), have been applied in various fields of technology for the last fifty years in parallel with worm-gear drives. They became widespread in drives of handling machinery, equipment and machines operating both at constant and variable loading modes.

Figure 1. The spiroid gear.

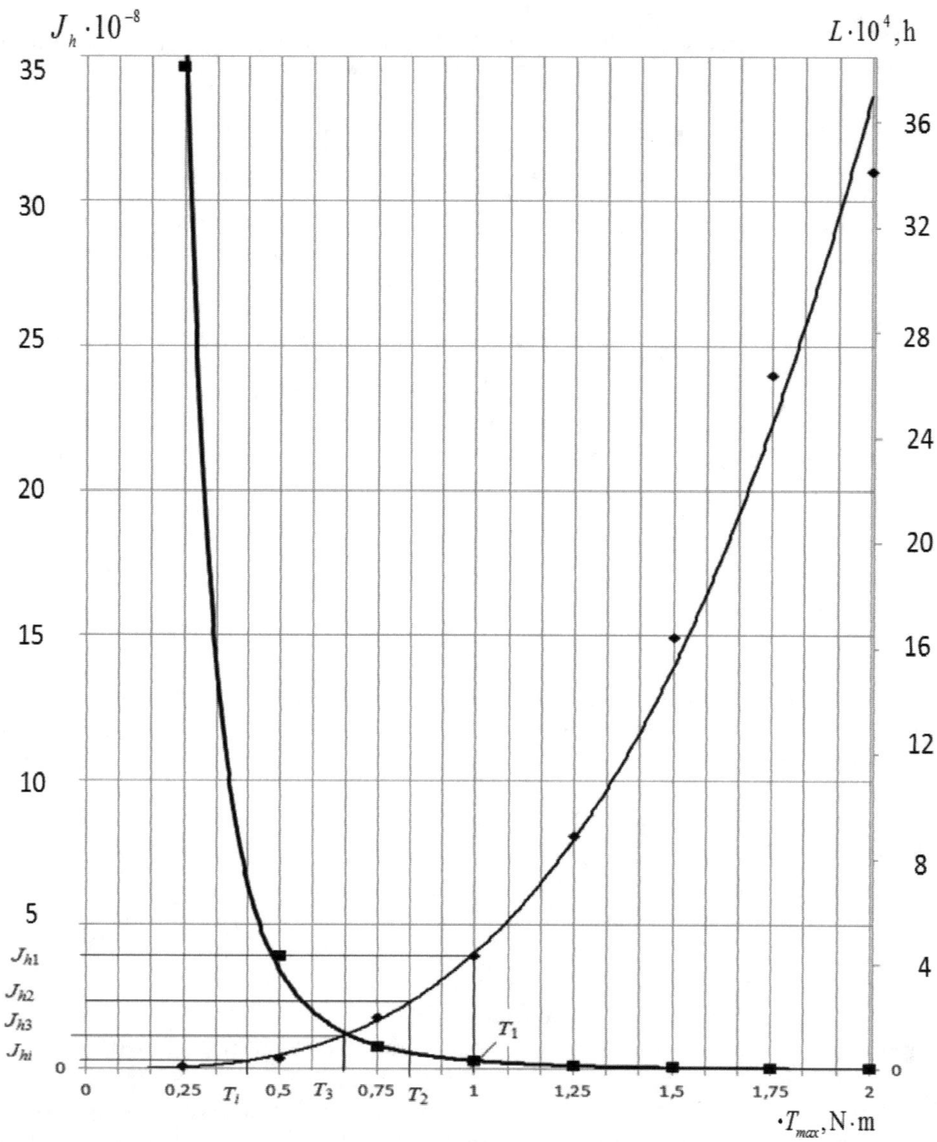

Figure 2. The procedure for determining the intensity of wear, depending on the values of torque on the output shaft of the spiroid gearbox according to the schedules of the wear and the resource intensity of the spiroid wheel on the values of torque on the output shaft of the spiroid gearbox.

IOP Conf. Series: Materials Science and Engineering **795** (2020) 012017 doi:10.1088/1757-899X/795/1/012017

Due to the fact that the considerable number of drives in mechanisms of lifting, construction, and road-building machines is operated at variable loading modes on the operating equipment during the cycle, creating a method of calculation of the spiroid resource by wear with regard of the loading variable nature becomes an actual challenge.

The complexity of building the universal method of calculating gears by wear is in determination of the values of driving torques on the gear drive output shaft during loading cycles and their duration with regard of operating conditions for lifting, construction, and road-building machines [9].

Moreover, the experimental determination of wear intensity of the cog-wheel material is of considerable labour intensity and complexity because it requires the creation of universal laboratory equipment for testing natural and model samples would provide the wide range for changing test conditions approximating the real modes of machines and mechanisms operation.

Figure 2 shows experimental dependencies of wear intensity J_h and resource L of the spiroid wheel cogs on the values of driving torque on the gear drive output shaft T_i. This dependence $J_h=F(T_i)$ allows finding wear intensity of the spiroid wheel cog for the corresponding driving torque values T_i of the set variable loading schedule.

2. Research methods.

Procedure of calculating the spiroid cylindrical transmission resource at variable loading mode.

1. Establish the mechanism loading schedule in accordance with the number of ranked relative values of drive torques on the spiroid gear drive output shaft in order of descending and with the relative duration of their action (see figure 3).

2. Calculate main parameters of the spiroid gear drive required for the resource calculation (Table 1).

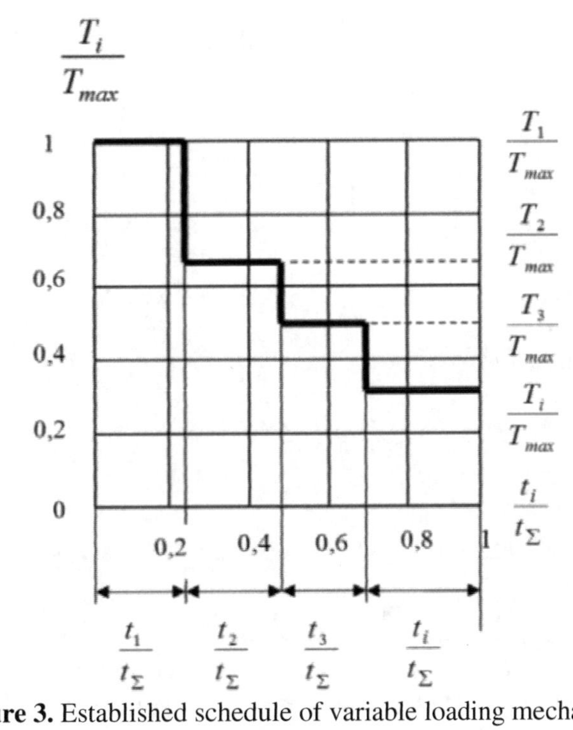

Figure 3. Established schedule of variable loading mechanism.

Table 1. The main parameters of the spiroid gearbox required to calculate the resource in accordance with GOST 22850-77.

№	Name of the main geometric parameters of the gearbox	Designation (calculation formula)	Dimension
1	Center distance	a_w	mm
2	Nominal (maximum) torque at the output shaft	$T_2=T_{max}$	N·mm
3	Nominal rotational speed of input shaft	n_1	min^{-1}
4	Nominal output speed	$n_2=n_1/u$	min^{-1}
5	Direction rotational speed of output shaft		
6	Efficiency coefficient in continuous operation (calculated)	η	
7	Gear ratio (from worm to worm wheel)	u_{12}	
8	Worm material	Steel 40Ch	
8.1	Modulus of elasticity	E_1	MPa
8.2	Poisson's ratio	μ_1	
9	Material of a spiroid wheel	Bronse Bra9Zh4	
9.1	Modulus of elasticity	E_2	MPa
9.2	Poisson's ratio	μ_2	
10	Roughness of worm turn surfaces	Ra	μm
11	Parameters of the worm		
11.1	Diameter factor of the worm	q	
11.2	Screw parameter	$P_y=mz_1/2$	mm
	Angle of inclination of the straight line	$\alpha=\gamma$	degree
11.3	Design module of the spiroid worm coil along the generatrix	m	mm
11.4	Calculated axial module of the spiroid worm	$m_x= m$	mm
11.5	Number of sets	z_1	
11.6	Dividing elevation angle of the worm turn line	γ	degree
11.7	Kind of worm		
11.8	Direction of the turn line		
11.9	Dividing diameter of the worm	$d_1=qm$	mm
11.10	Diameter of the turn tops	d_{a1}	mm
11.11	Length of the cut part	b_1	mm
11.12	Diameter of troughs	d_{f1}	mm
11.13	Dividing axial turn profile angle	α_{xR}	degree
		α_{xL}	degree
12	Worm wheel parameters		
12.1	Internal diameter	d_{i2}	mm
12.2	Outer diameter	d_{e2}	mm
12.3	Number of teeth wheel	z_2	
13	The given radius of curvature at the calculated point of contact of the wheel teeth		
13.1	For the right dividing axial angle of the turn profile	$(\rho_{red12})_R$	mm
13.2	For the left dividing axial angle of the turn profile	$(\rho_{red12})_L$	mm
14	Elastic constant of contiguous links	η	mm^2/N
15	Peripheral speed at calculated points of the wheel teeth profile	V_{FY2}	mm/s
16	Sliding speed at calculated points of the wheel teeth profile	V_{SY2}	mm/s
17	Lubricating material: CAT TDTO SAE 30 API GL-3 transmission oil - an analog of TM-3-9 according to Russian State Standard GOST 17479.2 - 85		

3. Calculate the maximum wear of the spiroid sheel cogs $[h_2]$ [10]:

$$[h_2] = [1{,}57 - (\mathrm{tg}\alpha_{xR} + \mathrm{tg}\alpha_{xL})]m_x . \tag{1}$$

4. Specific calculated force in the engagement corresponding to the driving torque on the gear drive output shaft T_i [10]:

$$w_{HEi} = \frac{T_i}{T_{max}} w_{HEmax}, \tag{2}$$

where T_i / T_{max}-relative driving torque values determined from the set variable loading schedule; w_{HEmax}-specific calculated force in the engagement corresponding to the driving torque on the gear drive output shaft T_{max}.

5. Establish the value of the gear drive spiroid wheel cog per the mechanism operating cycle, corresponding to the value of the operating driving torque on the gear drive output shaft T_i and its action duration t_i, mm [10]:

$$h_i = 2{,}25 J_{hi} \sqrt{w_{HEi}\eta(\rho_{red1,2})_R} \frac{V_{SY2}}{V_{FY2}} n_2 i\left(\frac{t_i}{60}\right), \tag{3}$$

where J_{hi}- the value of wear intensity of the spiroid wheel material, obtained from experiments, corresponding to the operating driving torque T_i (to be determined under the graph, see figure 2); t_i-duration of the operating driving torque T_i on the spiroid gear drive output shaft per its operating cycle in accordance with the set variable loading schedule (see figure 3), sec; i=1- umber of engagement couples with the cog-wheel under consideration.

6. Establish the value of the mechanism operation cycles to the maximum wear of the spiroid sheel cogs $[h_2]$ [10]:

$$N_\Sigma = \frac{[h_2]}{\sum_{i=1}^{n} h_i}, \tag{4}$$

where n- loading step number according to the set variable loading schedule.

7. Calculate the resource of the mechanism spiroid gear drive with the set variable loading schedule [10]:

$$L = \frac{N_\Sigma t_\Sigma}{3600} \tag{5}$$

3. Results and discussion

Dependence of the wear intensity J_h of the spiroid sheel cogs (material: bronze BrA9Zh4) on the values of the driving torque on the spiroid gear drive output shaft T_i was obtained from experiments, allowing to determine the wear intensity and calculate the resource of the mechanism spiroid gear drive with the set variable loading schedule.

4. Conclusions

The algorithm of calculating the resource of the spiroid gear drive by spiroid wheel wear based on the experimental dependance of the wear intensity on loading torque J_h=$F(T_i)$ is given.

IOP Conf. Series: Materials Science and Engineering **795** (2020) 012017 doi:10.1088/1757-899X/795/1/012017

Acknowledgments
Experimental dependence of the wear intensity J_h of the spiroid sheel cogs (material: bronze BrA9Zh4) on the values of the driving torque on the spiroid gear drive output shaft T_i was obtained based on bench test results of disc and roller analogy method.

References
[1] Drozdov Yu 2002 Gears and transmissions **2** pp 37-43
[2] Genkin M, Kuzmin N and Misharin Yu 1959 Gear jamming questions (USSR Academy of Sciences) p 147
[3] LtD Mechanic. Product Catalog http://mechanik.udmnet.ru/
[4] LtD Illinois tool works https://www.itw.com/
[5] Russian State Standard GOST 21354-87 (ST SEV 5744-86)
[6] Reshetov D, Gusenkov A, Drozdov Y et al 1995 Mechanical Engineering. Encyclopedia p 864
[7] Saari O 1960 Patent 2954704 USA. Skew-axis gearing
[8] Saari O 1954 Speed-Reduction Gearing, Patent USA №2696125
[9] Zaitsev A 2015 Mechanization of construction **10** pp 42-45
[10] Anferov V, Tkachuk A and Zaitsev A 2016 Fundamental and applied issues of mining sciences **3** 2 pp 11-17

IOP Conf. Series: Materials Science and Engineering **795** (2020) 012018 doi:10.1088/1757-899X/795/1/012018

The Development of Digital Technologies in Mining Machinery Technical Maintenance

V I Klishin, B L Gerike and A A Mokrushev

The Federal Research Centre of Coal and Coal Chemistry at the Siberian Branch of the Russian Academy of Sciences, prosp. Leningradskey 10, Kemerovo, 650065, Russia

E-mail: gbl_42@mail.ru

Abstract: Methods of functional diagnostics for technical conditions of rolling element bearings are studied in the paper. Their strengths and shortcomings are revealed. The model of shock pulses formation in the event of faults in the rolling element bearings is built. The possibility to apply wavelet-transformation for detecting these faults and defining technical condition of mining machinery assembly units is introduced.

1. Introduction

At present a great deal of time due to objectively reasonable causes industrial enterprises devote to the issues of improving the reliability, operating efficiency and maintenance of the processing equipment. These issues are vital, particularly, for hazardous production facilities such as coal and metal mining industry. The reason is in changes of approaches towards mining and conveyor equipment operation as the machinery and technological processes become more sophisticated, and the requirement towards industrial and environmental safety toughens. Great number of various machine modules have a concealed nature of faults origin and development generated within the period of their operation. This may cause accident events accompanied by significant economic and social damages and air pollution. The number of contemporary technogenic accidents and disasters of various scales [1] made it necessary to redefine the requirements towards reliable assessment of machinery condition and defining its residual operation life taking into account the latest scientific achievements in technical diagnostics [2, 3].

On the other hand, the majority of the enterprises under the condition of cost and budget cuttings come up against a dire necessity of decreasing the expenses. It also touches the issues of production modernization, repair and technical maintenance of main and auxiliary equipment as the share of operating equipment components and assemblies is in a worn-out state with significantly exhausted residual operation time level [4]. Under such condition, it is important that the cost minimization solution does not influence negatively on the machinery operation reliability. It is possible only if one can obtain accurate data on technical condition of the equipment and these data are obtained using different methods of technical diagnostics [5-9]. Any modern industrial enterprise pays close attention to improving its profitability by means of effective management of its business assets applying optimal strategy for repairing and technical maintenance. Russian and foreign industrial enterprise practices demonstrate that justified cost reduction on technical maintenance without decreasing its reliability is possible only with the help of complex application of modern sophisticated diagnostics methods.

At present, coal producers apply a significant number of extensible belt conveyers [10] and operating indices of the whole coal industry of Kuzbass depend on their good operating condition. In the short run, the increase of power availability and technical extensiveness of belt conveyers together with their performance and rock mass transportation distance is expected. One can see a large-scale application of variable-frequency electric drives [11, 12].

The increasing volumes of underground coal production require the creation of reliable transportation systems and it is the main issue, which the producers of coal mining extensible belt conveyer process lines should solve. Another, no less important issue is their technical maintenance costs reduction [13]. To provide faultless operation of the belt conveyer for the prolonged period of time the reasons which may cause the failure of different components [14] and especially of toothed gearings and bearing systems the life duration of which is judged by the sliding surface mechanical wear, should be defined.

2. Setting the task

The analyses of stand-by times caused by main and face conveyer reduction gear failures [15] shows that their share varies from 7 percent to 18 percent while the mean time for restoring normal operation makes it from 24 to 48 hours. All these prove the relevance of the researches.

Vibration monitoring method proved to be a reliable one for controlling technical condition of the mechanical equipment [5, 16, 17]. Vibration-based diagnostics is applied:

- for controlling current condition of the equipment;
- for dividing manifold admissible technical conditions of machines into two subsets - working and defective;
- for diagnosing that consists in defining the character and localization of one or a group of failures that correspond to the machine vibratory condition;
- for detecting possible failures at the early stage or predicting their temporal evolution;
- for estimating residual operation time;
- for defining repair time and volume;
- for reducing the risk of accidents.

The experience of monitoring technical condition of mining equipment indicates that to detect potential wear-out failures is more effective (to 77 percent) while applying the vibration parameters analysis[16] and if supported by other functional diagnostics methods such as oil spectral analysis [17] and thermal-imaging monitoring the accuracy of detecting the reason of faults increases to 95 percent.

Overall technical condition analysis of a reduction gear box after it has been assembled and tested by a testing facility (figure 1) will allow both detecting and isolating manufacturing faults and defects and preventing from delivering defective products to a consumer. Moreover, the obtained data can serve as a basis for the development of automated quality control system.

The analysis of vibration monitoring methods allows concluding that it is advantageous to apply spectrum-mask method (spectral plots) as a method for controlling output products of the coal industry. The idea of this method is in the fact, that the faults formed as a result of manufacturing and assembly works generate the vibration in specific frequency bands with definite magnitude relation of the controlled parameters.

Spectrum-mask method allows install the width of the frequency band, its position and evaluation criteria values which are compared with current values randomly. Analyzing the changes of the controlled parameter in a frequency band (the number of the bands can vary from 6 to 30) the evaluation and forecasting of the equipment condition is fulfilled [19].

Figure 1. General view on a testing facility at «Anzheromash Ltd».

The frequency ranges of a spectral mask (the band width) generally take the values relying on the following condition [5, 20]:

- «high-energy» components of the spectrum that accompany out-of-balance condition and misalignment – (0,5...1,5)×f_r and (1,5...2,5)×f_r;
- «low-energy» components of the vibration that accompany the rolling element bearings faults (defects) – (7,5...15,5)×f_r;
- (2,5...10,5)×f_r— general failure of a system rigidity;
- First mid-frequency band (3... 15)×f_r;
- Second mid-frequency band (15...40)×f_r;
- First high-frequency band 40×f_r...20 kHz;
- (n±1)×f_r – for failures in\of coupling boxes elements, where f_r – is a rotation frequency of a driving motor.

The captured data on all the types of the produced reduction gear boxes were statistically processed for every type of the reduction gear box in the form of maximum permissible level of a root mean square vibration speed [V_e] and spectral mask for each control point in three orthogonally related directions. For example, figure 2 presents the results of RKC-400 reduction gear box test run. It is done in the form of energy spectrum registered in axial direction in control point four.

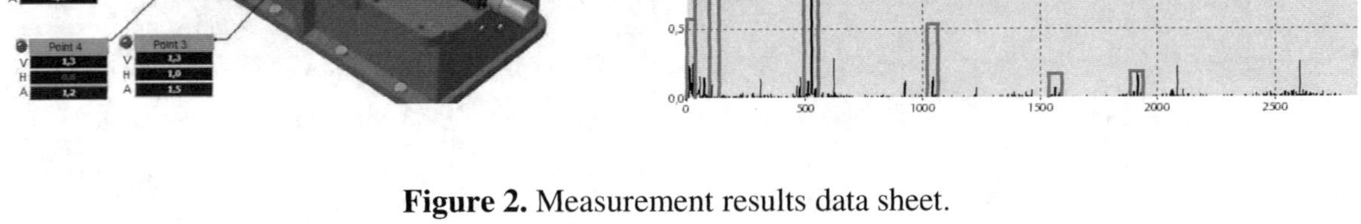

Figure 2. Measurement results data sheet.

Applying modern methods for machinery health monitoring allows practicing individual approach towards each manufactured machinery device while evaluating its technical condition and setting threshold values for its initial, functional and limit states.

Basic problems of mechanic faults in mining equipment (out-of-balance condition, misalignment, gear drive defects etc.) as a rule bring about the problems of bearing elements functioning in different power-driven, transforming and operating mechanisms of rolling element bearings.

At present the most comprehensive review on the possible rolling elements bearing defects is presented in [21, 22] which, according to the authors' opinion [21] is not complete and contains only basic faults, the reason that cause these faults and their localization.

The existing methods for analyzing technical condition of rolling elements bearings [2, 19, 23] in their rare cases, allow adequately detect the faults as the application of direct vibroacoustic signal spectral analysis for detecting the faults of the rolling elements bearings is hampered due to low amplitudes of their frequency components which are lost on the background of a "carpet noise".

3. Shock pulse modelling

To create more sensitive method for detecting faults and defects a detailed studying of the dynamic processes that take pace in rolling elements bearings and their modelling is necessary.

The initiation of a shock pulse can be described by the following model:

$$x(t) = a_0 \times \lambda^{-\beta t} \times \cos(\omega t + \varphi_0), \tag{1}$$

where x is a shift; a_0 is an initial amplitude; ω is the damped-vibration frequency, connected with the free-running frequency $\omega^2 = \omega_0^2 - \beta^2$; $\beta = rt/2m$ is vibration damping rate; r – resistance value; m – vibrating system mass.

Under the natural frequency system the frequency with which the system would vibrate without presence of resisting stress is understood [24]. An integral homogeneous system is studied in this model. In practice this process is more complicated as any machine consists of several parts (stator, rotor, frame, body tec.) which are movable towards each other. The energy from the shock is distributed between the assembly units making them vibrate with different, typical for them natural frequencies.

Figure 3*a* demonstrates the vibratory acceleration pulse shape registered on non-operating testing facility which was influenced by weak periodic stocks of a metal hammer. Even in the original signal (pulse) itself the presence of low and high frequency components is detected. Frequency components during Fourier transformation of vibroacoustic signal (figure 3b) that bring about false frequency components occurrence which are not typical for the original signal are mostly often detected.

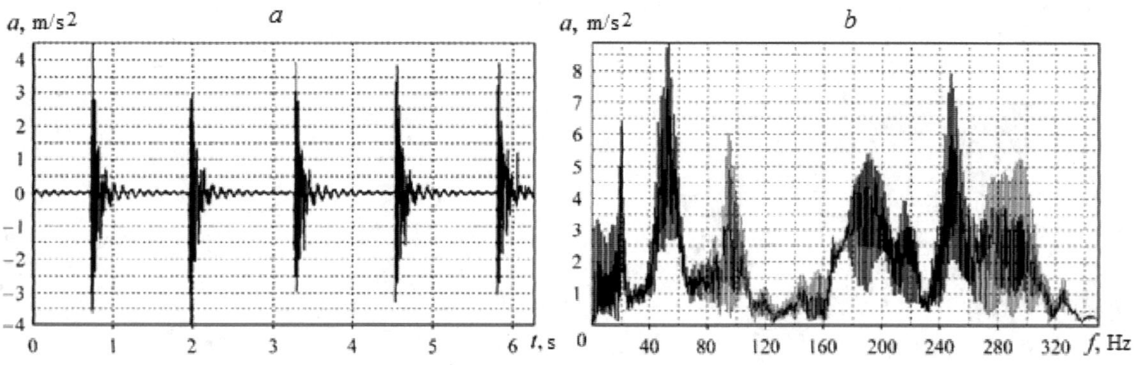

Figure 3. Vibratory acceleration signal (a) and its spectrum (b).

For each frequency different damping rates are typical. Figure 4 demonstrates the variation of different frequency component amplitudes of a central shock pulse in time located in the middle of a

IOP Conf. Series: Materials Science and Engineering **795** (2020) 012018 doi:10.1088/1757-899X/795/1/012018

timebase deflection. The presented decays are divided into equal time intervals. The analogy with electromagnetic wave can be traced as depending on the length (frequency) they have larger or lesser property of being absorbed (dissipate). On the other hand, it is connected with the mass of an assembly unit that creates the given frequency and with the presence of any energy absorber (for example, shock absorber).

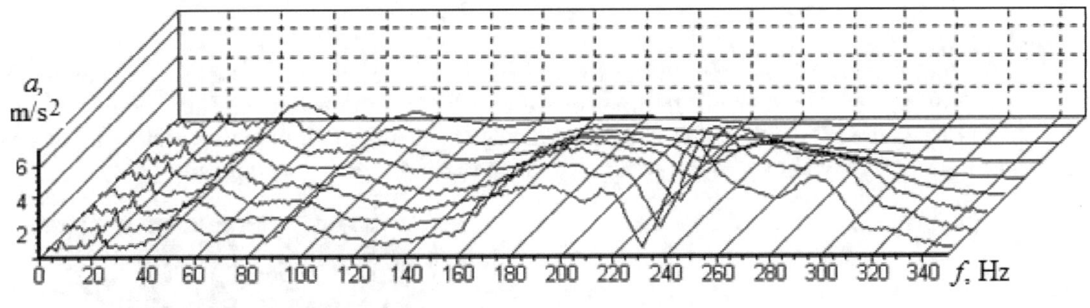

Figure 4. The development of the pulse frequency in time.

The difference in a shock pulse amplitude at the initial moment of time is conditioned by the geometry of a mechanism, i.e. by the distance between the source of the impulse and a specific assembly unit, presence of impediments for dissemination of a vibroacoustic signal. Supposing that the studied system is linear the real pulse we will represent as a sum of pattern pulses with different frequencies and damping rates:

$$I(t) = \sum_{i=1}^{N} a_i \times \lambda^{-\beta_i t} \times \cos(\omega_i t + \varphi_i). \qquad (2)$$

To transform the original signal into the Eq. (1) it is optimal to expand it on the bases $x_i(t)$, but the system $x_i(t)$ is nonorthogonal. Consequently it is necessary to build orthonormal system $\{[e_1(t), e_2(t)], ..., [e_i(t), e_j(t)], \{[e_i(t), e_j(t)] = 0, \forall i \neq j, \forall i \in 1...N\}$ by analogy with Fourier's series, along with minimal correction of the original system $x_i(t)$.

In seventies of the twentieth century wavelet methods appeared. Two limits are imposed on a wavelet function W_Ψ: it should be isolated enough i.e. go to zero when distancing from the origin of coordinates; the integral of the function at $(-\infty; +\infty)$ should be equal to zero. Wavelet transformation looks as:

$$W_\Psi f(a,b) = \frac{1}{\sqrt{C_\Psi}} \int_{-\infty}^{\infty} \frac{1}{\sqrt{|a|}} \Psi\left(\frac{b-x}{a}\right) f(x)dx. \qquad (3)$$

where a is a scale; b is a shift; $C_\Psi = 2\pi \int_{-\infty}^{\infty} (|\Psi(\omega)|^2 / |\omega|) d < \infty$ – normalizing constant, $\Psi(\omega)$ is Wavelet W_Ψ Fourier transform. Freedom in selecting basis functions $\Psi[(b-x)/a]$ allowed introducing many types of wavelets (Haar, Daubechies, Gaussian, Morlet etc.)

Wavelet transform is a signal of homomorphous and short wavelets, which can be shifted or stretched along the time axis. This is a fundamental difference from infinite wave of Fourier transforms [25, 26].

Apart from continuous wavelet transforms there is a discrete transformation where the filtration process takes place. Due to all these, the notions of approximation (high-scale high frequency

IOP Conf. Series: Materials Science and Engineering **795** (2020) 012018 doi:10.1088/1757-899X/795/1/012018

components) and components (low-scale one) occur. As a result it appears that the original signal is divided into two signals and they complement each other giving two time larger amount of information about the original one. In comparison with the signal decomposition into Fourier-series the wavelets can present local peculiarities of signals with better accuracy and solve the problem of detecting the faults and defects in the equipment by a complex way.

4. Modal testing

Wavelet function has all essential properties for solving the above-mentioned task. For example, the development of the pulse frequencies in time (figure 5) is built with the help of Haar modified wavelet decomposition. As it was mentioned above the advantage of this method is in its isolation i.e. it gives the chance to trace the dynamics of frequency components amplitudes development.

a *b*

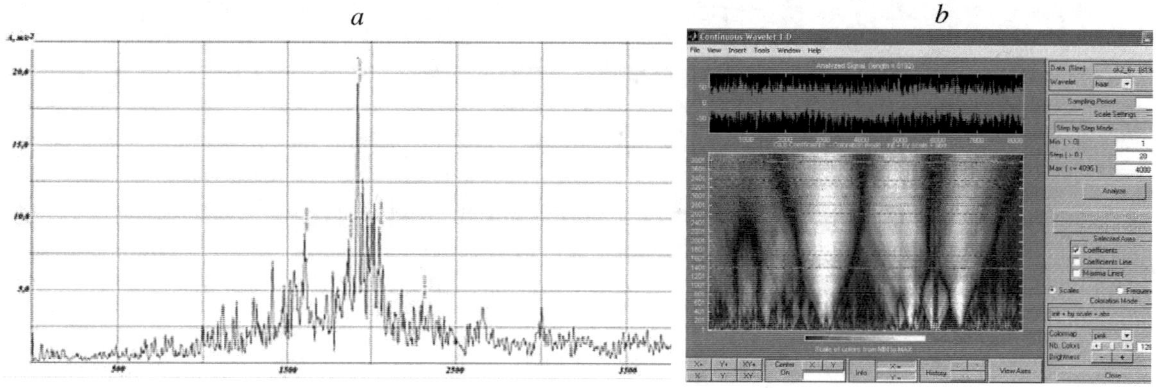

Figure 5. Fourier decomposition (a) and Haar wavelet decomposition (b) of vibroacoustic signal of the developed rolling element bearing fault.

Figure 5 shows the result of vibroacoustic signal decomposition measured in control point 5 and introduced in the form of energy spectrum (figure 5*a*) and Haar welvet (figure 5*b*). The analysis of the data shows that the vibro-activity in control point 5 is over the permissible standards.

a *b*

Figure 6. Double-row bearer defect:
a) the wear-out of the rolling bearing body; b) the wear-out of a bearing track.

5. Conclusion

The introduced approach for rating the mechanical vibration parameters can be used in practice while developing the industrial standards for norming the output product vibration for the purpose of enclosing it into the datasheet.

The development of the large number of spectral masks for the wide-range of mining machinery is one of the conditions for quality product manufacturing by mining machinery factories and it assists in shifting towards new forms of technical maintenance of mining machinery.

Applying the forecasting model based on statistic data of the vibration-based diagnostics give a chance adequately evaluate the researched fault and forecast the residual life of an assembly unit or machinery making maintenance planning more effective and preventing the occurrence of emergency failures.

In general the introduced solution will allow minimizing expenses connected with sudden failures of rolling element bearings, optimizing supply logistics and storage facility. All the conditions will be created for shifting to a brand new system of mining machinery repairing and technical maintenance.

References

[1] The Decree of the President of RF dated on December, 31, 2015 # 683 On the Strategy of National Security of the Russian Federation

[2] Non destructive testing: reference book 2005 vol 7 **B2** under the editorship of Klyuyeva V V (Moscow Mashinostroyeniye) p 829

[3] Sushko A Ye and Gribanov V A 2011 The problem of evaluation of dynamic machinery technical condition of dangerous industrial objects *Industrial safety* **10** 58–65

[4] Kovalyev V A, Khoreshok A A and Gerike B L 2015 Technical diagnostics of roadheader of selective action *Ugol* **9** 42–7

[5] Kelly S Graham 2013 *Advanced vibration analysis* (Dekker mechanical engineering) p 637

[6] Gilyev A V, Tchesnokov V T, Lavrov N B etc. 2011 *Mining machinery operational activity* (Krasnoyarsk: Sibirskey Federal University) p 276

[7] Kostyukov V N and Naumenko A P 2014 *Basis of vibroaoustic diagnostics and monitoring* (Novosibirsk: SB RAN Publishing house) p 329

[8] Kravchenko V M, Sidorov V A, Sedush V Ya 2009 *Technical dyagnostics of mechanical machinery* (Donetsk: Yugo-Vostok) p 459

[9] Gerike B L, Gerike P B, Kvaginidze V S etc. 2012 *Mining machinery diagnostics* (Moscow: IPO U Nikitskikh vorot) p 400

[10] Gammershmidt A A 2015 Conditions and perspectives of Kuzbass coal and carbon industry development *Ugol* **5** 14–5

[11] Galkin V I and Sheshko Ye Ye 2010 *Transport machines* (Moscow: Gornaya kniga) p 588

[12] Franco Jefferds dos Santos Silva, Herbert Ricardo Garcia Viana and André Nasser Aquino Queiroz 2016 Availability forecast of mining equipment *Journal of Quality in Maintenance Engineering* vol 22 (4) pp 418 – 432

[13] Peton N. and Drygin S 2017 Elevated vibration of a 110 MW generator bearing *SIRM 2017, 12th International Conference on Vibrations in Rotating Machines* (Graz, Austria)

[14] Gajanand Gupta and Rajesh P Mishra 2016 A SWOT analysis of reliability centered maintenance framework *Journal of Quality in Maintenance Engineering* vol 22 (2) pp 130 – 145

[15] Anil Rana 2016 Optimal maintenance level of equipment with multiple components *Journal of Quality in Maintenance Engineering* vol 22 (2) pp 180 – 187

[16] Krakovskey Yu M 2005 *Mathematic and programme software for evaluating technical condition of the equipment* (Novosibirsk, Nauka) p 200

[17] Popaleny P, Peton N and Walker G 2019 Cryogenic pumps monitoring, diagnostics and expert systems using motor current signature analyses and vibration analyses *Proceedings of the ASME-JSME-KSME 2019 Joint Fluids Engineering Conference* (San Francisco, CA, USA)

[18] Klyuyev V V, Vavilov V P etc. 2005 *Non-destructive testing: reference book Thermal control* (Moscow, Mashinostroyeniye) 9–361

[19] Peton N, Drygin S, Leonov D 2017 Resolving structural vibration issues on a water flood pump *45th Turbomachinery Symposium* (Houston, Texas, USA)

[20] Preventive maintenance of equipment 1990 (CSI) p 252
[21] Rolling bearings defects diagnostics 2009 (NSK Motion & Control) p 42
[22] Rolling bearings defects and the reasons causing these defects 2002 (SKF AB) p 46
[23] Rudloff L, Arghir M, Bonneau O, Guingo S, Chemla G and Renard E 2012 Experimental analysis of the dynamic characteristics of a hybrid aerostatic bearing *J. Engineering for Gas Turbines and Power* **134** 18
[24] Astafyev N M 1996 Wavelet analysis: theoretical basis and application samples *Uspekhi Phis. Nauki* **166** 11 1145–70
[25] Vityazev V V 2001 *Wavelet analysis of time series* (St.Peterburg SPbSU) p 58
[26] Dyakonov V P 2012 *Matlab:self-study book* (St. Peterburg DMK press) p 770

IOP Conf. Series: Materials Science and Engineering **795** (2020) 012019 doi:10.1088/1757-899X/795/1/012019

Electromagnetic Method for Determining the Internal Resistance of Conductive Materials

F Y Kuznetsov

National Research Tomsk Polytechnic University, Tomsk, Russia

kuznetsov_f@mail.ru

Abstract. The electromagnetic method of vibration diagnostics of rod conductive parts is theoretically described. Natural frequencies of the transverse oscillations of the rod are determined for various boundary conditions. In theory, method of experimental determination of internal friction coefficient of electrically conductive materials is developed.

1. Introduction

It is known that oscillations caused in a solid body fade relatively quickly even in the absence of external resistance. For a long time, this process was not given due importance. Kelvin was the first to notice this phenomenon. A follower of his research was Voigt, who marked the beginning of a microscopic theory. The first review of the problem of internal friction, aimed at clarifying the essence of this phenomenon, was made by K. Ziener [1,2].

Internal friction is understood as the phenomenon of dispersion of energy of elastic oscillations due to internal processes in a solid body, during which mechanical energy of oscillations is converted into thermal energy [1].

With the development of technology, electromechanical systems are everywhere present in our lives. But they are often exposed to vibrations. In their design and manufacture it is important to accurately determine the damping factor, and it primarily depends on the internal friction. Therefore, the experimental determination of the internal friction coefficient is an urgent problem.

There are four hypotheses describing the phenomenon of fading oscillations in an elastic body due to internal friction. The most common is the Voigt hypothesis, according to which the internal friction forces during oscillations are proportional to the speed of movement (viscous friction). The Voigt model has some drawbacks, for example, it is not able to display the relaxation of the material, as compared with the Maxwell and Poiting-Thomson models, which are close to the properties of real bodies [2].

Methods of measurement of internal friction can be divided into the following classes:

- method of free fluctuations;
- resonance method;
- waves propagation method;
- direct determination of stress-strain curves.

Application of the first three classes of methods is possible with some assumptions about the nature of dissipative forces and the linearity of the system. These assumptions are the dissipative force is proportional to the rate of change of deformation (Voigt model) and the type of mechanical behavior does not depend on the deformation amplitude in the stress region used in the experiments. In our case, the measurement of internal friction was carried out using the resonance method.

The electromagnetic method of vibration diagnostics is based on using an external magnetic field. The presence of the magnetic field leads to a distribution of electromagnetic force that acts on the rod. In the case of natural oscillations, the magnetic force is fading, which leads to a change in the attenuation factors of the partial oscillations. Electromagnetic influence possesses selectivity with respect to oscillatory modes, thus, the attenuation factors of various partial oscillations of the rod vary to different degrees. Therefore, it is possible to determine the modes whose factors do not change when the rod vibrates in the magnetic field [3].

The purpose of this study is the theoretical substantiation of the electromagnetic method for determining the internal friction coefficient of conductive materials and its experimental confirmation.

There are several causes of internal friction in materials, and these causes can be recognized by their natural frequency. The task of determining the coefficient of internal friction is important for materials science.

2. Resonant method of determination of internal friction coefficient

An experimental test of the electromagnetic method for determining the internal friction coefficient was carried out on a specially designed installation. A stand is mounted on a massive bed on which an metallic rod 40 cm long and 6 mm in diameter is mounted on a cantilever arm. Aluminum and copper were chosen as the rod material. An external uniform magnetic field is created by two flat magnets, which are fixed so that the induction vector is directed perpendicular to the plane of oscillation of the rod. With the help of a special form waveform generator Tabor WW2571A, a specific current of frequency is passed through the rod. There are forced vibrations of the rod under the action of electromagnetic force. When the frequency of the driving force coincides with any natural frequency of the rod oscillations, a sharp increase in the amplitude of the oscillations occurs - a resonance.

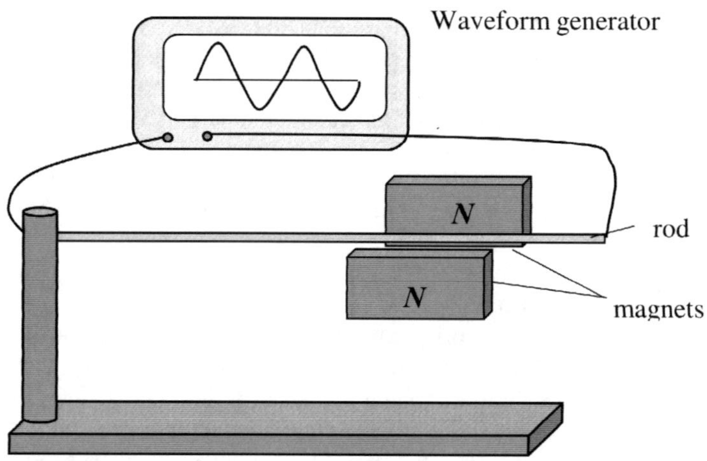

Figure 1. Experimental installation for natural frequencies of fluctuation determination.

In the paper by Tomilin A.K. [3]. The process of transverse oscillations of an electrically conductive rod in an external magnetic field is theoretically described. Based on the data obtained in the monograph [3], we can draw the following conclusions:

- the set of frequencies of an electrical signal depends on the width and location of the active region, on which an external magnetic field acts;
- the electromagnetic effect has a selective property, i.e. a group of natural frequencies of vibrations of the rod, which are not reflected in the electrical signal, is distinguished;
- the magnetic force acting on the rod increases the fading factor of free oscillations, therefore in each particular case it is necessary to compare the magnitude of the mechanical and electromagnetic dissipation.

To determine the internal friction with transverse oscillations of a cantilever rod in a magnetic field using the resonance method, we write the equation of transverse oscillations of a rod in a magnetic field, taking into account the exciting electromagnetic force [3]:

$$EJ\left(\frac{\partial^4 u}{\partial z^4}+\beta^*\frac{\partial^5 u}{\partial z^4\partial t}\right)+m_0\left(\frac{\partial^2 u}{\partial t^2}+\beta\frac{\partial u}{\partial t}\right)+\frac{B^2}{R}\int_{z_1}^{z_2}\frac{\partial u}{\partial t}dz=F_{эм}(t)\ , \tag{1}$$

where: $u(z,t)$ - the dynamic function of displacements, depending on the longitudinal coordinate and time; EJ - flexural rigidity of the rod; β - the coefficient of external damping depending on properties of the environment; β^* - the damping coefficient depending on the properties of the material of the rod; m_0 - linear mass of the rod; *R*- full resistance of an electrical circuit. The last member in the (1) corresponds to electromagnetic force.

Let's present the function of the driving force in the form:

$$F_{эм}(t)=\sum F_{0i}\sin\omega_i t\ .$$

Let's transform the integral-differential (1) to the system of the ordinary differential (2), using the Fourier procedure and the condition of orthogonality of amplitude functions:

$$\ddot{q}_r+\left(\beta+\beta^* p_r^{\ 2}\right)\dot{q}_r+p_r^{\ 2}q_r=-\frac{B^2\gamma_r}{\chi_r m_0 R}\sum_{n=0}^{\infty}\dot{q}_n\gamma_n+\frac{j_B AB\gamma_r}{\chi_r m_0},\qquad (r=1,2,...) \tag{2}$$

where: q_r - the generalized coordinate; B - magnetic induction; j_B - current density; A - core cross-sectional area; $\chi_r=\int_{z_1}^{z_2}X^2{}_r dz$; $\gamma_r=\int_{z_1}^{z_2}X_r dz$.

From (2), we determine the damping factor $\tilde{h}_r$ and the damped frequency $\tilde{p}_r$ of the first oscillations, using the formula:

$$\tilde{h}_r=\frac{1}{2}\left(\beta+\beta^* p_r^{\ 2}+\frac{B^2\gamma_r^{\ 2}}{\chi_r m_0 R}\right);\qquad \tilde{p}_r=\sqrt{p_r^{\ 2}-\tilde{h}_r^{\ 2}}\ . \tag{3}$$

The external resistance (air resistance) and the magnetic dissipation of oscillations have small values compared with the internal friction of the material, therefore, approximately it is possible to accept:

$$\tilde{h}_r=\frac{1}{2}\beta^* p_r^{\ 2}\ . \tag{4}$$

The maximum amplitude of oscillations in the experiment was achieved at the frequency of the electric signal $\omega=23{,}0\pm0.1s^{-1}$ (for aluminum) and $\omega=14.7\pm0.1s^{-1}$ (for copper). With a further increase in the frequency of the input signal, the amplitude of oscillations of the rod gradually decreased. Accordingly, the cyclic damped frequency has the meaning:

$$\tilde{p}_{alum}\approx144.4s^{-1}\ ;\ \tilde{p}_{copper}\approx92.31s^{-1}.$$

Having determined the experimental cyclic frequency $\tilde{p}$, through (3) we can find the coefficient of internal friction β^*:

$$\beta^*=2\frac{\sqrt{p_r^{\ 2}-\tilde{p}_r}}{p_r^{\ 2}} \tag{5}$$

Table 1 shows the calculations of the internal friction coefficient, attenuation factors and damped frequencies during the induction of magnets *B = 1 T* and the action of the magnetic field in the area with coordinates $z_1=0.7l$, $z_2=0.85l$. The network impedance is 10^6 Ω.

Table 1. The experimental data

Number of oscillations	Material	$p_r\left(s^{-1}\right)$	$\omega\left(s^{-1}\right)$	$\beta^*\left(s^{-1}\right)$	$\tilde{h}_r\left(s^{-1}\right)$
r=1	Aluminum	167.51	144.4	0.0061	84.89
r=1	Copper	115.87	92.31	0.0104	70.04

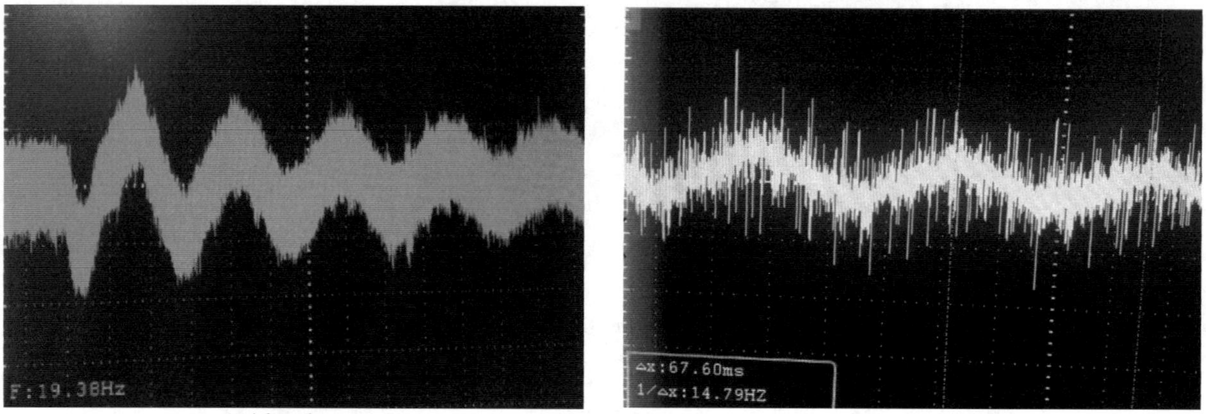

a) Aluminum b) Copper

Figure 2. Forced oscillation frequency.

Figure 2 shows the graph and the value of the forced oscillation frequency of the rod used to verify the correct determination of the natural vibration frequency by the resonance method. Comparing the oscillograms for aluminium (figure 2a) and copper (figure 2b), it is possible to replace that the value of damping decrement and damping factor of aluminium exceeds the value of damping decrement and damping factor of copper, which is confirmed by calculations and shown in Table 1.

3. Conclusion

The results obtained in this study show possibilities of the electromagnetic method. It can be used to experimentally determine the natural frequencies of transverse oscillations of a rod-shaped object and determination of coefficient of internal friction of the materials from which they are made.

The advantages of this method are: the absence of damage to the test sample, the non-contact measurement method, the invariability of the mechanical properties of the sample during the test. The advantages include a small amount of magnetic dissipation, compared with the impedance of the frequency analyzer. This means that natural frequencies of the electromagnetic method of vibration diagnostics coincide with the frequencies without exposure to a magnetic field.

References

[1] Blanter M S 2004 What internal friction is about *Sorosovskiy obrazovatel'nyy zhurnal* **8(1)** pp 80-85
[2] Blanter M S, Golovin I S and Neuhauser H 2007 H-R-Sinning Internal Friction in Metallic Materials *Springer* 539 p
[3] Tomilin A K and Baizakova G A 2012 Control over oscillation frequencies of resilient electromechanical system *Vestnik Tomskogo gosudarstvennogo universiteta. Matematika i mekhanika* **3(19)** pp 87-92
[4] Tomilin A K and Baizakova G A 2012 Electromagnetic method of adjusting the frequency of the vibrometer *Izvestiya vuzov. Fizika* **6/2** pp 244-247.

[5] Tomilin A K and Kurilskaya N F 2017 Vibrations of a conductive string in a nonstationary magnetic field under presence of two nonlinear factors *Journal of Applied and Industrial Mathematics* **11(4)** pp 600–604 DOI 10.1134/S1990478917040184
[6] Lavrovich N I 2000 Own vibration frequencies of rods *Omsk Scientific Herald* pp 106-108
[7] Ziyatkhan Aliyev and Sevinc Guliyeva 2017 Properties of natural frequencies and harmonic bending vibrations of a rod at one end of which is concentrated inertial load *Journal of Differential Equations* **263(9)** DOI: 10.1016/j.jde.2017.07.002
[8] Steinhauser J and Nad M, 2015, Applied Mechanics and Materials, **808** 315-320
[9] Lin Du, Weijuan Li, Shengshi Zhao and Shansheng Wang 2017 Internal Friction Study of Aging Hardening and Kinetics in Low Carbon Steel *International Conference on the Technology of Plasticity (ICTP)* **17-22** (*Cambridge, United Kingdom*)
[10] L B Magalas 2009 Mechanical spectroscopy, internal friction and ultrasonic attenuation: Collection of works *Materials Science and Engineering* vol 521–522 pp 405-415

IOP Conf. Series: Materials Science and Engineering **795** (2020) 012020 doi:10.1088/1757-899X/795/1/012020

Load Capacity of Spherical Roller Transmission with Double-Row Pinion

M E Lustenkov and E S Lustenkova

Belarusian-Russian University, Mira Avenue 43, Mogilev, 212000, Republic of Belarus

E-mail: lustenkov@yandex.ru

Abstract. This paper provides an in-depth description of the design and operating principle of a spherical roller transmission producing a wide range of reduction ratios in small dimensions. The rotational motion is transmitted from the drive shaft to the driven shaft by a pinion with two coaxial rows of rollers. The pinion performs a spherical motion about a fixed point on the transmission axis with a constant nutation angle. The rollers are in contact with a fixed periodic closed groove formed by two cams mounted in the housing and a groove made on the cam of the driven shaft. Power is transmitted through a number of parallel flows, which increases the load capacity, i.e., the maximum torque in the driven shaft. The main criterion of reliability is the contact strength of the cam surfaces. Equations for calculating contact stress in the roller-groove contact on the cam of the driven shaft are obtained by use of the Hertz theory. A formula for calculating the maximum torque being transmitted, which is obtained from the kinetostatic equilibrium equations, is given. The analysis of the influence of the main geometric parameters of the transmission on its load capacity is presented.

1. Introduction

Mechanical transmissions with high reduction ratios are used in space and aeronautical systems, drives of positioning and servo systems, shut-off devices, etc. The use of cascade-connected gear sets leads to the greater weight and larger dimensions of the drive, a higher level of noise, increased backlash and, as a result, reduces the reliability of the mechanical system. To obtain high speed reduction with small dimensions, different types of mechanical transmissions are used: worm, wave, planetary gears, etc. Reduction ratios of planetary transmissions with a double-gear pinion can be more than 10,000 [1], but their efficiency is very low. Pin-cycloid gears have a higher efficiency [2], but in both cases the pinion mounted on the drive shaft with eccentricity needs to be balanced. Precession (nutation) transmissions, whose pinion performs a spherical motion about a fixed point [3] (its center of mass), are more balanced. Precession drives with bevel wheels are characterized by low noise and higher load capacity [4].

Spherical roller transmissions (SRT) with a double-row pinion are innovative mechanisms; they are similar in the structure and the principle of operation to precession transmissions. SRTs belong to the class of spherical mechanisms characterized by balance due to the central symmetry, the ability to simultaneously perform several functions (for example, a speed reducer and a spherical joint), as well as by additional kinematic capabilities [5, 6]. SRTs can produce a wide range of reduction ratios, as in the case of planetary transmission with a double-gear pinion. Due to small dimensions, light weight, coaxial alignment of shafts and cylindrical (or spherical) housing, they can perform functions of unified mechanical modules and can be integrated into kinematic chains of robots and manipulators. It has been found that the main criterion of reliability of these transmissions is contact strength [7]. It determines the SRT load capacity, the maximum torque being transmitted through the driven shaft. The purpose of this study is to conduct a theoretical assessment of the influence of the main geometric parameters on the SRT load capacity.

2. Design and principle of operation of SRT

Figure 1 shows SRT drive shaft 1 with a section inclined to the transmission axis at an angle Θ. Pinion 2 with the capability to rotate relative to the section is mounted on it. The pinion has two coaxial rows of holes. The rollers with projecting spherical ends are put in these holes with uniform angular spacing. When drive shaft 1 rotates, rollers of the outer row 4 are in contact with periodic groove 7 made in housing 8. Rollers of the inner row 3 are in contact with the groove of end cam 5. Cam 5 is mounted on driven shaft 6 and rotates with it. The groove in the housing is made by two end cams, displaced relative to each other by a quarter period of the groove as shown in figure 2. Both grooves have a rounded profile in the normal section and are produced on cams by means of a spherical mill; the center of the spherical mill, like the center of the spherical end of the rollers, moves along the closed periodic curve located on the spherical surface. The inclination of the pinion axis to the axis of the drive shaft is produced by means of a bushing with a hole, which is inclined at an angle to the axis of the outer surface of the bushing, and two taper washers.

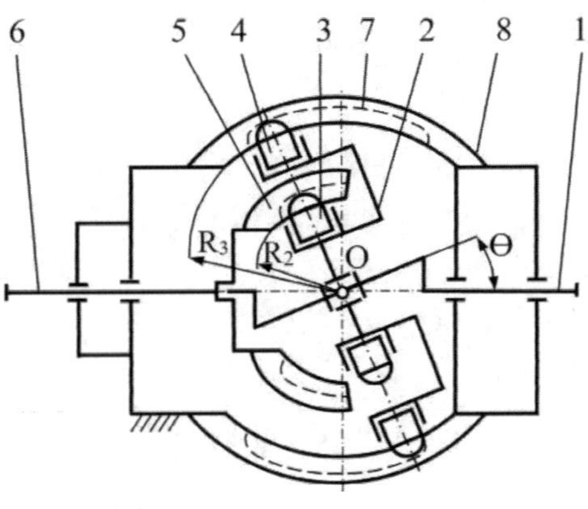

Figure 1. Kinematic diagram of SRT: 1 – drive shaft; 2 – pinion; 3 – rollers of the inner row; 4 – rollers of the outer row; 5 – end cam; 6 – driven shaft; 7 – fixed periodic groove; 8 – housing.

Figure 2. Key components of SRT prototype: 1 – drive shaft with the double-row pinion; 2 – driven shaft with the end cam; 3 – end cams forming the fixed periodic groove.

The main geometric parameters of SRT are radii of spheres R_2 and R_3, on which the centers of the spherical ends of the rollers of the inner and outer rows are located, respectively, and the angle of the drive shaft section Θ. The number of periods of the housing groove Z_3 and periodic working surface of the end cam determines the reduction ratio i. It can be expressed as:

$$i = Z_2(Z_3+1)(Z_2-Z_3)^{-1}. \qquad (1)$$

The number of the rollers of the inner row n_{s1} and the outer row n_{s2} is one more than the corresponding number of the periods Z_2 and Z_3.

The center of mass of the pinion is located on the transmission axis, which improves its balance; dynamic pressures on the pinion bearings are insignificant compared with the forces in engagement. The increase in load capacity is produced by a larger number of rollers simultaneously involved in, engagement and transmitting load. The rollers are put into the pinion holes by means of bushings made from antifriction material. The rollers are capable of rolling on the groove surface, as in transmissions with intermediate rolling bodies, without sliding. [8–10]. But, in contrast to these transmissions, rollers axes are fixed on a common support, the pinion, which helps reduce noise and avoid jamming.

3. Foundations of SRT contact strength calculations

The condition of contact strength in relation to SRT is obtained by transforming the Hertz equation for the area of contact between the ball and the curved groove [11]:

$$\sigma_H = K_{Gs}\xi_q N_{m\max}^{\frac{1}{3}} \sum \rho^{\frac{2}{3}} \le [\sigma_H], \tag{2}$$

where σ_H, $[\sigma_H]$ are the calculated actual and permissible contact stresses, respectively, MPa; $N_{m\max}$ is the maximum value of the normal reaction in the roller-groove contact, N; K_{Gs} is the factor depending on the properties of contacting materials, $MPa^{2/3}$; ξ_q is the coefficient depending on the factor Ω that in turn depends on differences in the curvature of contacting surfaces. The factor Ω for the contact area of the roller spherical surface and the groove can be expressed as:

$$\Omega = \left(\frac{1}{r_s + \Delta r_s} - \frac{1}{R_2 + r_s}\right)\left(\frac{2}{r_s} - \frac{1}{r_s + \Delta r_s} - \frac{1}{R + r_s}\right)^{-1}, \tag{3}$$

where r_s is the radius of the spherical surface of the roller, mm; Δr_s is the value by which the radius of the groove profile exceeds the radius r_s, mm.

The factor ξ_q can be determined from tables [12] taking into account data interpolation. To automate the calculations, the data from these tables were processed, a dependency graph was created and a 7th-degree polynomial was obtained. It is expressed as:

$$\xi_q = 1.02 - 0.83\Omega + 11.25\Omega^2 - 67.47\Omega^3 + 197.58\Omega^4 - 304.13\Omega^5 + 235.21\Omega^6 - 72.22\Omega^7. \tag{4}$$

With the materials of the parts being the same, the contact stresses in the materials of the cams are determined, because the roller surface is, as a rule, harder. The factor K_{Gs} is calculated according to the following expression:

$$K_{Gs} = 0.364\left(\left(1-\mu_r^2\right)E_r^{-2} + \left(1-\mu_t^2\right)E_t^{-2}\right)^{-\frac{2}{3}}, \tag{5}$$

where $\mu_{r(t)}$ is the Poisson's ratio, which depends on the properties of the materials of the rollers (r) and the groove (t); $E_{r(t)}$ is the modulus of elasticity, MPa.

The maximum value of normal reactions will be achieved in the contact area between the inner row rollers and the cam mounted on the driven shaft, since $R_2 < R_3$, and only half of the rollers can be engaged at the same time. To determine the average value of the normal reaction N_{2m}, a mathematical model of the force interaction of the rollers with the cams surfaces was developed by the kinetostatic method. Thus, we obtain the following expression:

$$N_{2m} = \frac{T_2 K_{nsf} \cos\psi}{0.707 K_n K_{p2}\left(R_2 + 0.707 r_s\right) n_{s2} \sin\left(\alpha_{2m} + \psi\right)}, \tag{6}$$

where T_2 is the nominal torque on the driven shaft, Nm; K_{nsf} is the factor used to determine mean values of the radii of the circles formed by the portions of the spherical surface cut off by planes perpendicular to the transmission axis; ψ is the friction angle, rad; K_n is the factor describing uneven distribution of load on the rollers; K_{p2} is the ratio of the number of rollers transmitting the load to their total number indicating the number of the rollers transmitting the load to their total number; α_{2m} is the average inclination of the groove on the cam of the driven shaft, rad.

The friction angle corresponds to the reduced friction coefficient f: $\psi = \text{atan} f$. The average groove inclination is determined on the plane unfolding of the spherical curve when replacing it with a piecewise helical curve (the vertices of the periodic curve are connected by line segments). It is expressed as $\alpha_{2m} = \text{atan}(2Z_2\Theta\pi^{-1})$.

The factors included in the expression (6) can be calculated as follows:

$$K_{Nsf} = 0.25\left(1+\cos\Theta\right)^{-1}. \tag{7}$$

$$K_{p2} = \left(1-0.707 r_s \tan\alpha_{m2}\cdot\sin\alpha_{m2}\left(\Theta\cdot R_2\right)^{-1}\right)Z_{s2}, \tag{8}$$

where Z_{s2} is the factor depending on the type of closure of kinematic pairs formed by the rollers and the groove. With self-closed kinematic pairs, if the groove is formed by two cams, $Z_{s2} = 1.0$. With force-closed kinematic pairs, the rollers are in contact with one cam, and the number of rollers simultaneously transmitting the load is reduced by half. Thus, $Z_{s2} = 0.5$.

Substituting formula (6) into expression (2), and replacing the computed contact stresses σ_H with their permissible values $[\sigma_H]$, we obtain the equation for the torque T_2, which is the maximum torque that can be transmitted $T_{2\max}$ based on the contact strength criterion:

$$T_{2\max} = S\frac{\left[\sigma_H\right]^3 K_n K_{p2}\left(R_2+0,707r_s\right)n_{s2}\sin\left(\alpha_{2m}+\psi\right)}{\xi_q^3\Sigma\rho^2 K_{nsf}\cos\psi}, \tag{9}$$

where $S = 1.098\cdot10^{12}$ is the factor taking into account the dimensions of the parameters in formula (9).

4. Evaluation of the influence of SRT parameters on the load capacity

The main geometric parameter of SRT which affects the value of the torque being transmitted and its efficiency is the groove inclination α_{2m}. This inclination, in turn, depends on three independent parameters: the radius R_2, the angle Θ (or the groove amplitude $A_2 = R_2\Theta$) and the number of periods Z_2 related to the reduction ratio i.

Let us consider the transmission whose cams and rollers are made from AISI 5115 steel (DIN 16MnCr5) with working surface hardened to 60...62 HRC by carburizing. In this case, the yield strength is $\sigma_T = 930$ MPa, the permissible contact stresses in prolonged operation is $[\sigma_H] = 1150$ MPa. For steel parts $E_r = E_t = 2.1\cdot10^5$ MPa, $\mu_r = \mu_t = 0.3$. With the materials of rollers and grooves being the same, $K_{Gs} = 863.6$ MPa$^{2/3}$. The numbers of periods are $Z_2 = 15$, $Z_3 = 13$, the reduction ratio is $i = 105$. The radius of the spherical end of the roller is assumed to be equal to the radius of the groove profile in the normal section, i.e., $\Delta r_s = 0$ mm.

The results of calculations with fixed $R_2 = 38$ mm are shown in figure 3. The values of the experimental factor $K_n = 0.9$, which corresponds to parts accuracy class 7 and the reduced friction coefficient $f = 0.02$, which takes into account the rollers' rolling and sliding [13], were assumed to be constant. Thus, the study of the influence of geometric parameters of SRT on its load capacity at the specified radial dimensions and reduction ratio has been performed.

Theoretically, an increase in r_s should cause an increase in the torque being transmitted, because the radii of curvature of the contacting surfaces and the contact areas increase as well, and contact stresses are replaced by bearing stresses. There is an optimal value of the radius r_s based on the criterion of maximum load capacity. There are some factors that have the most significant influence. When the groove inclination α_{2m} decreases, which is equivalent to a decrease in Θ and Z_2 and an increase in R_2, the radii r_s, which produce the extrema $T_{2\max}$, increase, and, as a result, the values of the torques increase. The influence of the angle Θ (and, respectively, α_{2m}) with a stepwise change in r_s can be observed in figure 4. It should be noted that there are optimal values of α_{2m} (and Θ) based on the maximum efficiency criterion. Small groove inclinations cause an increase in wedging forces; at certain values this can lead to SRT's self-braking and jamming.

Another important factor affecting the SRT load capacity is the ratio K_{p2}. When the radius r_s increases, the working sections of the groove-roller contacting areas are reduced.

There is also a limit on the maximum value of the radius based on the smooth operation criterion. The maximum limit of the radius can be calculated as follows:

$$r_{s\max} = \left(1-\frac{1}{Z_{s2}n_{s2}}\right)\frac{R_2\Theta}{0.707\tan\alpha_{2m}\sin\alpha_{2m}}. \tag{10}$$

When this value is exceeded, the average number of the rollers simultaneously transmitting the load becomes less than one, which leads to failure of the engagement smoothness.

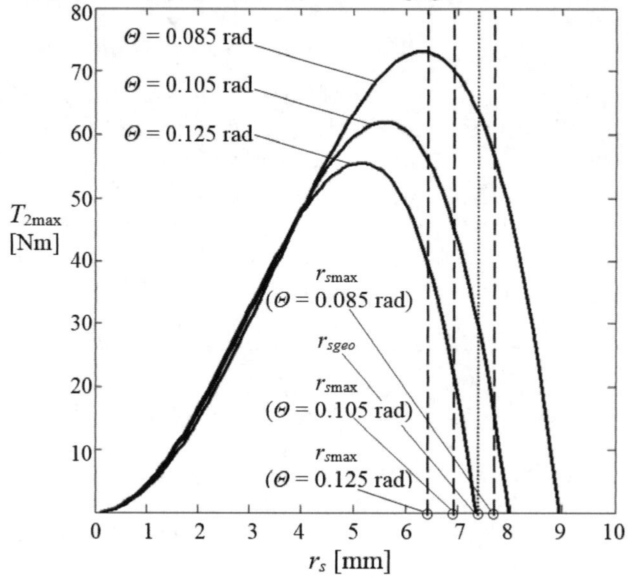

Figure 3. SRT load capacity dependence on the radius of the roller spherical end.

Another restriction is related to arranging rollers in the SRT pinion by analogy with the planet gear arrangement in the planetary gear. This geometric condition of avoiding interference on the roller surfaces can be expressed by the maximum value of the radius r_{sgeo} corresponding to the specified R_2 and n_{s2}: $r_{sgeo} = R_2\sin(\pi/n_{s2})$.

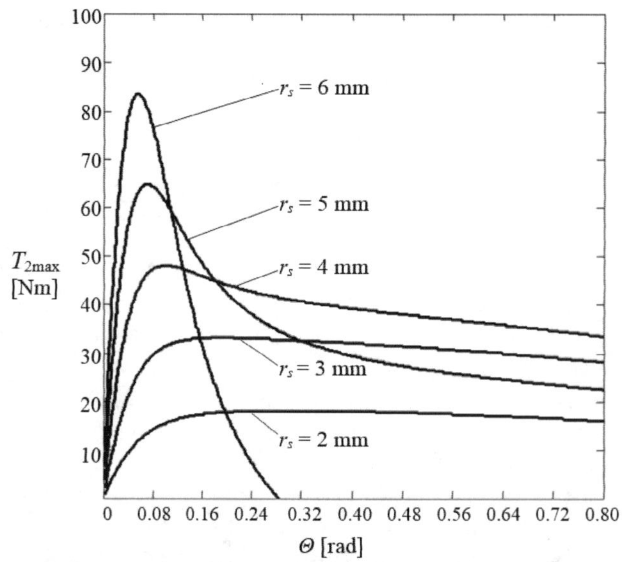

Figure 4. SRT load capacity dependence on the angle Θ.

Table 1 presents the maximum torque values T_{2max} for each Θ and respective values r_s, as well as the restrictions imposed on the radius value r_s.In addition to the efficiency, the shear stresses acting in the roller cross section perpendicular to its axis and the bearing stresses arising in the contact area between the roller and the pinion hole must be taken into account.

Table 1. Influence of SRT parameters on the load capacity.

Θ (rad)	$T_{2\max}$ (Nm)	r_s (mm)	$r_{s\max}$ (mm)	r_{sgeo} (mm)
0.075	81.993	6.8	8.457	7.413
0.085	73.120	6.3	7.814	7.413
0.095	66.743	5.9	7.329	7.413
0.105	61.994	5.6	6.995	7.413
0.115	58.358	5.3	6.661	7.413
0.125	55.506	5.1	6.424	7.413
0.135	52.231	5.0	6.232	7.413

5. Conclusions

A mathematical model has been developed, which makes it possible to calculate the main geometric parameters of SRT with specified radial dimensions and reduction ratio and achieve the maximum load capacity taking into account computed contact stresses. To increase the load capacity, the radius r_s of the spherical ends of the rollers should be increased and the groove inclination α_{2m} should be reduced. When the radius r_s increases, the maximum torque being transmitted increases as well, but the range of the angles Θ being used is narrowed down. Finally, the value of α_{2m} must be adopted taking into consideration its influence on the SRT efficiency. An increase in the radius R_2, with the other parameters being constant, leads to an increase in the radial dimensions of SRT, a decrease in α_{2m} and, consequently, an increase in its load capacity. When estimating the influence of the reduction ratio i, we must take into account the fact that it depends on the difference in the numbers of the groove periods $Z_2 - Z_3$. In any case, an increase in the number of the periods Z_2 results in an increased α_{2m} and a decreased SRT load capacity determined according to the contact strength criterion.

References

[1] Sclater N 2011 *Mechanisms and Mechanical Devises* (McGraw-Hill)

[2] Meng Y, Wu C and Ling L 2007 Mathematical modeling of the transmission performance of 2K–H pin cycloid planetary mechanism *Mechanism and Machine Theory* vol 42 iss 7 pp 776–790

[3] Molyneux W G 1997 The internal bevel gear and its applications *Proc. Inst. Mech. Eng.* G **211** pp 39–61

[4] Hong J, Yao L, Ji W and Huang Z 2015 Modeling for the Nutation Drive Based on Screw Theory *Procedia CIRP* **36** pp 123–128

[5] Bai S and Angeles J 2009 The Design of Spherical Multilobe-cam Mechanisms *Proc. IMechE* vol 223 part C: *J. Mechanical Engineering Science* pp 473–482

[6] Zhang L, Pan C, Xu X, Xu H and Zhang Z 2016 An investigation of cam-roller mechanism applied in sphere cam engine Journal of Central South University **23**(4) pp 825–833

[7] Lustenkov M E 2010 Planetary Ball Transmissions: Strength Calculations *Russian Engineering Research* **30**(9) pp 862–866

[8] Wang G X and Guan T M 2009 Modeling of Nutation Drive with Rolling Teeth *Applied Mechanics and Materials* **16–19** pp 708–712

[9] Prudnikov A P 2018 Thermal analysis of transmission with intermediate rolling bodies *AER-Advances in Engineering Research* (*AviaENT* 2018) **158** pp 338–342

[10] Nam W K, Shin J W and Oh S H 2013 Design of thin plate-type speed reducers using balls for robots *Journal of Mechanical Science and Technology* **27**(2) pp 519–524

[11] Johnson K L 1987 *Contact Mechanics* (Cambridge: Cambridge University Press)

[12] Cooper D H 1969 Hertzian contact-stress deformation coefficients *Trans. ASME, J. Appl. Mech.* **36** pp 296–3022

[13] Efremenkov E A and I-Kan A 2010 Euler-Savari determination of radii of curvature of cycloid profiles *Russian Engineering Research* **30**(10) pp 1001–04

Identification of Mathematical Model of Drying Unit

F P Perevoshchikov, N V Vuong, A K Shabo and A A Shilin

Tomsk Polytechnic University, 30, Lenina ave., Tomsk, 634050, Russia

E-mail: fvp2@tpu.ru

Abstract. In air drainage systems, the process of reducing moisture content is based on its cooling down to the dew point and the precipitation of liquid into the condensate. Under isobaric cooling, the dew point temperature depends on the relative humidity and temperature in the drained space, which in turn can be controlled by the temperature of the heater and the air flow rate created by the fan, the main part of which is the electric drive.

1. Introduction

Thus, due to the fact that the system is two-dimensional, there is some uncertainty in the control of the heater and the fan motor, due to the functional dependence of the transmission coefficient on the power of the heater and the fan motor respectively. Consequently in a modern world, accurate and fast positional control is required.

One of the solutions to this problem is to obtain an analytical solution from the equations of heat conduction and thermodynamics, taking into count the formula for saturated vapor pressure derived by Magnus in 1844 [1]:

$$P_{wp_i}(t^{\circ}) = a \cdot exp\left[\frac{b \cdot t^{\circ}}{c + t^{\circ}}\right] \tag{1}$$

where a,b,c - empirical coefficients varying from model to model [1]. There are many formulas built on the general definition of Magnus (Table 1). The most abundant and used in meteorology models of Abott-Teboni, Alduchov, and Buck can be distinguished [2-4]. Non-Magnus models obtained by Wexler [5] are also standing apart.

But this approach has drawbacks related to the fact that for complex multicomponent and multicriteria systems, obtaining exact analytical solutions is not a trivial task. And even if this succeeds, then most often the mathematical model of the system is inadequate.

Table 1. Basic formulas for saturated vapor pressure above water.

Model Author	Abbreviation	Formula (kPa)
Abott-Teboni	AT85	$P_{wp_i}(t^{\circ}) = 0.61070 \cdot exp\left[\frac{17.38 \cdot t^{\circ}}{239.0 + t^{\circ}}\right]$
Alduchov	AERK	$P_{wp_i}(t^{\circ}) = 0.61094 \cdot exp\left[\frac{17.625 \cdot t^{\circ}}{243.04 + t^{\circ}}\right]$
Buck	BU81	$P_{wp_i}(t^{\circ}) = 0.61121 \cdot exp\left[\frac{17.502 \cdot t^{\circ}}{240.97 + t^{\circ}}\right]$
Wexler	WE76	$P_{wp_i}(T) = 0.001 \cdot exp[-2.9912729 \cdot 10^{3} \cdot T^{-2} - 6.0170128 \cdot 10^{3} \cdot T^{-1} + +1.887643845 \cdot 10 - 2.8354721 \cdot 10^{-2} \cdot T + 1.7838301 \cdot 10^{-5} \cdot T^{2}] - -8.4150417 \cdot 10^{-10} \cdot T^{3} + 4.4412543 \cdot 10^{-13} \cdot T^{4} + 2.858487 \cdot lnT]$

2. Dryer Model

Therefore, a prototype was created for modeling processes in drying plants (Figure 1, a – inside, b - outside). The prototype was based on a plastic box, an Arduino Nano microcontroller, a block of resistors that act as a heater, a Peltier element, the cold junction of which is in the box and two fans, one of which creates the necessary air flow in the drained volume, and the second removes heat from the element's hot junction Peltier.

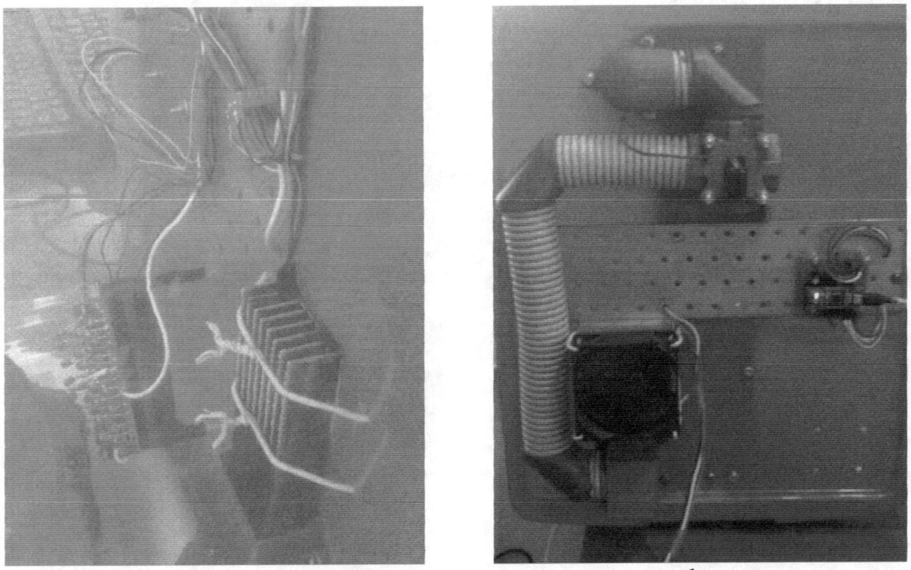

a b

Figure 1. Dryer prototype.

During the experiment, the static modes were removed and a rectangular temperature grid was obtained on the cooling radiator and in the box for a number of capacities consumed by the heater and the electric fan: 20.40.60% (Figure 2, a – $t^{\circ}_{rad} = t^{\circ}_{rad}(P_{heat}, P_{fan})$, b - $t^{\circ}_{box} = t^{\circ}_{box}(P_{heat}, P_{fan})$, Table 2). The temperature transient curves were also obtained when the power consumption of the fan motor and the heater changed (Figure 3, a – $t^{\circ}_{rad}(t)$, b - $t^{\circ}_{box}(t)$, Figure 4, a – $t^{\circ}_{rad}(t)$, b - $t^{\circ}_{box}(t)$,).

Table 2. Experimental temperature data.

P_{heat}, %	P_{fan}, %	t°_{rad}, $^{\circ}C$	t°_{box}, $^{\circ}C$
20	20	7.625	29.125
20	20	8.813	28.875
20	20	9.500	28.875
40	40	11.500	35.875
40	40	12.375	35.375
40	40	13.250	35.250
60	60	13.750	42.000
60	60	13.750	40.375
60	60	15.375	39.563

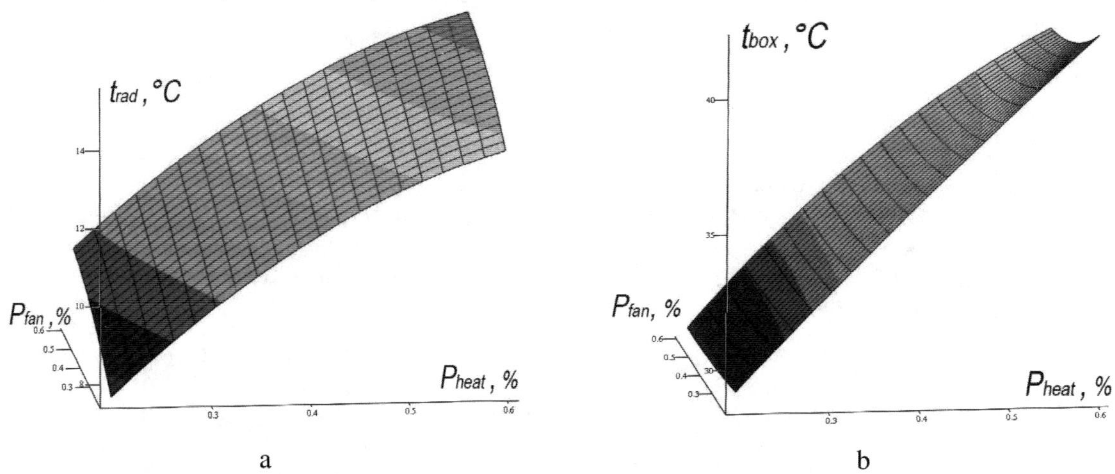

Figure 2. Experimental surfaces $t^{\circ}_{rad} = t^{\circ}_{rad}(P_{heat}, P_{fan})$ and $t^{\circ}_{box} = t^{\circ}_{box}(P_{heat}, P_{fan})$.

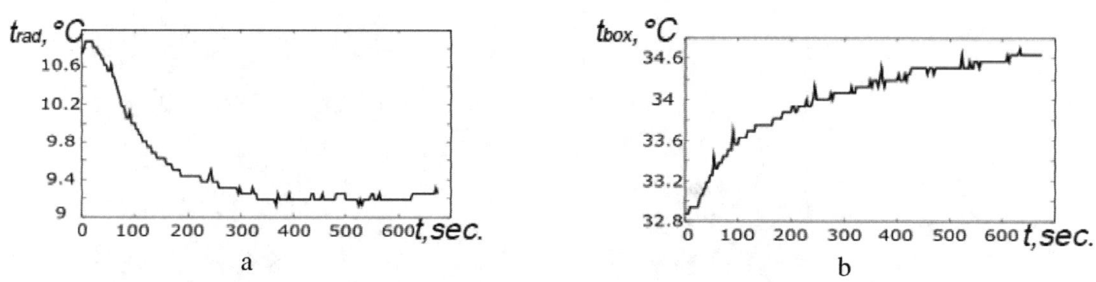

Figure 3. Transient responses $t^{\circ}_{rad}(t)$ and $t^{\circ}_{box}(t)$ while decreasing fan power consumption.

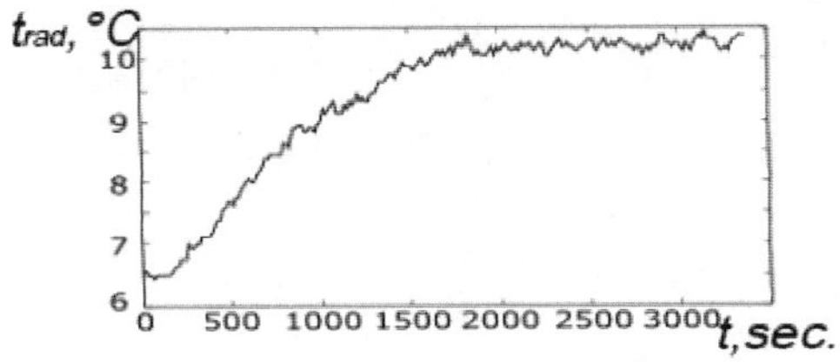

a

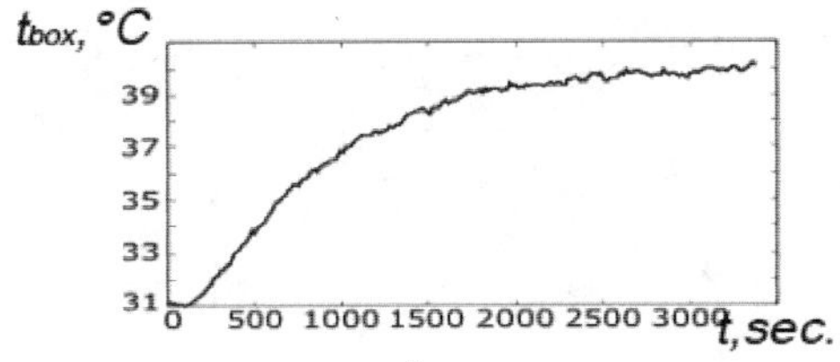

b

Figure 4. Transient responses $t^{\circ}_{rad}(t)$ and $t^{\circ}_{box}(t)$ with increasing heater power consumption.

3. Identification of mathematical model

To get features $t^{\circ}_{rad} = t^{\circ}_{rad}(P_{heat}, P_{fan})$, $t^{\circ}_{box} = t^{\circ}_{box}(P_{heat}, P_{fan})$ we will use numerical methods involving the use of a rectangular grid $t^{\circ}_{rad.ij} = t^{\circ}_{rad}(P_{heat.ij}, P_{fan.ij})$ where $i = \overline{1,3}; j = \overline{1,3}$ Due to the small number of nodes, it is reasonable to use the well-known method of the interpolating Langrazh polynomial:

$$t^{\circ}_{rad/box}(P_{heat}, P_{fan}) = \sum_{i=1}^{3}\sum_{j=1}^{3} t^{\circ}_{rad/box.i,j} \cdot l_{i,j}(P_{heat}, P_{fan}) \quad (2)$$

where $l_{i,j}(P_{heat}, P_{fan})$ - basic Lagrange polynomials, which are defined as:

$$l_{i,j}(P_{heat}, P_{fan}) = \prod_{p=1, p\neq i}^{3} \prod_{q=1, p\neq j}^{3} \frac{(P_{heat} - P_{p.heat})(P_{fan} - P_{q.fan})}{(P_{i.heat} - P_{p.heat})(P_{j.fan} - P_{q.fan})} \quad (3)$$

According to experimental data, using (2) using the Mathcad software environment, dependencies were determined $t^{\circ}_{rad}(P_{rad}, P_{fan})$ and $t^{\circ}_{box}(P_{rad}, P_{fan})$:

$$\begin{aligned} t^{\circ}_{rad}(P_{heat}, P_{fan}) &= 125 \cdot P_{heat}^{2} \cdot P_{fan} - 156.25 \cdot P_{heat}^{2} \cdot P_{fan}^{2} - 39.0625 \cdot P_{heat}^{2} - 1.875 \\ &+ 125 \cdot P_{heat} \cdot P_{fan}^{2} - 101.5625 \cdot P_{heat} \cdot P_{fan} + 46.875 \cdot P_{heat} - 25 \cdot P_{fan}^{2} + 25 \cdot P_{fan} \end{aligned} \quad (4)$$

$$\begin{aligned} t^{\circ}_{box}(P_{heat}, P_{fan}) &= 24.2188 \cdot P_{heat}^{2} \cdot P_{fan}^{2} - 69.2189 \cdot P_{heat}^{2} \cdot P_{fan} + 5.063 \cdot P_{heat}^{2} - \\ &-0.7813 \cdot P_{heat} \cdot P_{fan}^{2} + 27.0313 \cdot P_{heat} \cdot P_{fan} + 33.063 \cdot P_{heat} + 1.125 \cdot P_{fan}^{2} - \\ &-5.05 \cdot P_{fan} + 22.715 \end{aligned} \quad (5)$$

As a result, nonlinear functions were obtained $t^{\circ}_{rad}(P_{rad}, P_{fan})$ and $t^{\circ}_{box}(P_{rad}, P_{fan})$, which should be linearized by expanding in a Taylor series before the first amendment to find the transmission coefficient:

$$\begin{aligned} t^{\circ}_{rad}(P_{heat}, P_{fan}) &= \left.\frac{\partial t^{\circ}_{rad}}{\partial P_{heat}}\right|_{ij} P_{heat} + \left.\frac{\partial t^{\circ}_{rad}}{\partial P_{fan}}\right|_{ij} P_{fan} = K_{heat.ij} P_{heat} + K_{fan.ij} P_{fan} = (-78{,}125 P_{heat} - \\ &-101{,}5625 P_{fan} - 312{,}5 P_{heat} \cdot P_{fan}^{2} + 125 P_{fan}^{2} + 250 P_{heat} \cdot P_{fan} + 46{,}875)|_{ij} \cdot P_{heat} + \\ &(-101{,}5625 P_{heat} - 50 P_{fan} - 312{,}5 P_{heat}^{2} \cdot P_{fan} + 125 P_{heat}^{2} + 250 P_{heat} \cdot P_{fan} + 25)|_{ij} P_{fan}; \end{aligned} \quad (6)$$

$$t^{\circ}_{box}(P_{heat},P_{fan}) = \left.\frac{\partial t^{\circ}_{box}}{\partial P_{heat}}\right|_{ij} P_{heat} + \left.\frac{\partial t^{\circ}_{box}}{\partial P_{fan}}\right|_{ij} P_{fan} = K_{heat.ij}P_{heat} + K_{fan.ij}P_{fan} = (10{,}125P_{heat}$$

$$+27{,}0313P_{fan} + 48{,}4375P_{fan}{}^{2}P_{heat} - 0{,}78125P_{fan}{}^{2} - 138{,}4375P_{fan}\cdot P_{heat} + \tag{7}$$

$$+33{,}0625)|_{ij}\cdot P_{heat} + (27{,}0313P_{heat} + 2{,}25P_{fan} + 48{,}4375P_{fan}P_{heat}{}^{2} - 69{,}21875P_{heat}{}^{2} -$$

$$-1{,}5625P_{fan}P_{heat} - 5{,}05)|_{ij}\cdot P_{fan}$$

And in the end, it became possible to write linear differential equations for the heat exchanger radiator temperature and the box temperature using the obtained transmission coefficients contained in functions (6) and (7):

$$\dot{t}^{\circ}_{rad}(t) = \frac{(K_{heat.ij}P_{heat} + K_{fan.ij}P_{fan}) - t^{\circ}_{rad}(t)}{\tau_{inertion}} \tag{8}$$

$$\dot{t}^{\circ}_{box}(t) = \frac{(K_{heat.ij}P_{heat} + K_{fan.ij}P_{fan}) - t^{\circ}_{box}(t)}{\tau_{inertion}} \tag{9}$$

4. Conclusion

Thus, complex nonlinear dynamic processes in drying plants can be represented by a linearized mathematical model. In the assembled prototype of the installation based on the obtained interpolants, we can conclude that the radiator temperature is well controlled by changing the power consumption of the fan motor and heater, and the box temperature can be controlled by the heater, but not by the fan due to nonmonotonicity $t^{\circ}_{rad}(P_{fan})$. But these measurements are not the initial data for the synthesis of control systems of other dryers. The result of this work is a method of obtaining $t^{\circ}_{rad}(P_{heat},P_{fan})$ and $t^{\circ}_{box}(P_{heat},P_{fan})$ for the purpose of writing and solving differential equations $F(t,t^{\circ}_{rad}(t),\dot{t}^{\circ}_{rad}(t))=0$ and $F(t,t^{\circ}_{box}(t),\dot{t}^{\circ}_{box}(t))=0$ only on a certain grid of temperature values, which will allow us to synthesize a controller based on a linearized model to establish the required quality of work.

References

[1] Magnus G 1844 Versuche uber die spannkrafte des wasserdampfes Annalen der Physik, **137(2)** 225-247
[2] Abbott P F and Tabony R C 1985 The estimation of humidity parameters Meteorological Magazine **114(1351)** 49-56, Alduchov O A and Eskridge R E 1996 Improved Magnus form approximation of saturation vapor pressure Journal of Applied Meteorology **35(4)** 601-609
[3] Buck A L 1981 New equations for computing vapor pressure and enhancement factor Journal of Applied Meteorology **20(12)** 1527-1532
[4] Wexler A 1976 Vapor pressure formulation for water in the range 0 to 100 J.Res. Nat. Bur. Stand page 775

IOP Conf. Series: Materials Science and Engineering **795** (2020) 012022 doi:10.1088/1757-899X/795/1/012022

Planetary Torus Helical Transmission

A P Prudnikov

Belarusian-Russian university, Mira Ave, 43, Mogilev, Republic of Belarus

E-mail: prudnikovap85@gmail.com

Abstract. The paper describes design considerations for the transmission with fixed intermediate rolling bodies with grooves cut in a torus. Equations of helices used for cutting grooves have been derived. The transmission with the proposed design has a wide range of reduction ratios and small dimensions with high load capacity.

1. Introduction

Nowadays, many industries need drive systems with small dimensions and high load capacity. This is usually the case in drilling equipment, power-driven devices, industrial robots and mechatronic modules. In the drives of these mechanisms, planetary gears and harmonic drives as well as transmissions with intermediate rolling bodies, which were developed at the end of the 20th century, are mostly used [1].

Planetary gears offer high load capacity and small dimensions with reduction ratios from 1 to 20. The disadvantage of harmonic drives with reduction ratios from 80 to 300 is the presence of a flexspline that reduces its efficiency and load capacity. Transmissions with intermediate rolling bodies have a wide range of reduction ratios (from 3 to 200) with high load capacity due to distribution of load among a number of rolling bodies.

Disadvantages of transmissions with intermediate rolling bodies include low efficiency and reliability caused by the fact that, during operation of the transmission, rolling bodies move in the slits of the driven shaft. To increase efficiency and load capacity of the transmission, it is advisable to use rollers as rolling bodies. Rollers forms translational kinematic pairs with the slits and the grooves they interacts with, and unavoidable clearance, which occurs in the engagement, leads to the roller skewing when the load is applied. This causes sliding friction to occur in the contact area of rollers with parts of the transmission and can result in its failure. In transmissions with intermediate rolling bodies, there is a relationship between the reduction ratio and load capacity. With an increase in the reduction ratio due to cutting the vertices of the multiperiodic groove, the number of rolling bodies transmitting the load decreases (with reduction ratios of more than 12 the force is transmitted through less than 50 % of the number of rollers). In this case, with an increase in the reduction ratio, the total number of rollers in the transmission increases, and their dimensions decrease (due to the constraint imposed on dimensions of the transmission by its functional purpose). Hence, for transmissions with intermediate rolling bodies, the rational range of reduction ratios providing high efficiency and load capacity is from 3 to 30.

2. Design and principle of operation of the transmission

The solution to the problem of the roller skew is to create a transmission with fixed intermediate rolling bodies [2], i.e., when rolling bodies and the driven shaft form a kinematic pair with rotational and not translational motion. Based on this principle, a design of a planetary torus helical transmission with fixed intermediate rolling bodies (composite rollers) has been proposed. This transmission is called a torus transmission since the grooves are made on a torus. In this case, cylindrical and radial configurations of the transmission can be obtained.

In the cylindrical configuration, the equation of a helix located on the torus along the axis of rotation of the driven shaft is used for obtaining grooves. In the radial configuration, the helix (Archimedean spiral) is located on the torus perpendicular to the axis of rotation of the driven shaft (figure 1).

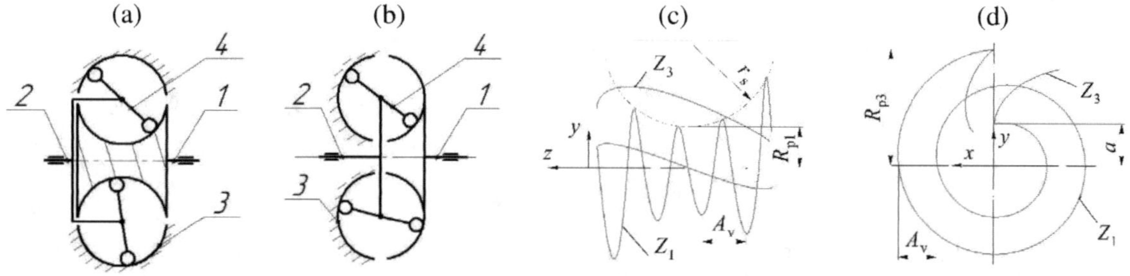

Figure 1. Cylindrical (a) and radial (b) configurations of planetary torus helical transmission and helices (c), (d) forming grooves: 1 – drive shaft; 2 – driven shaft; 3 – fixed member; 4 – rolling bodies.

Like the classic transmission with intermediate rolling bodies, the planetary torus helical transmission, as shown in figure 1, consists of four main parts: drive shaft (1), driven shaft (2), fixed member (3) and rolling bodies (4). The drive shaft is made with an open groove which has a shape of a helix (Archimedean spiral) with a whole number of periods Z_1 (in figure 1 (c) $Z_1 = 4$, in figure 1 (d) $Z_1 = 2$). The fixed member has open grooves with quantity of Z_3 (a whole number), which have the shape of an incomplete helix (Archimedean spiral) cut with equal pitch on the torus of the fixed member. In figure 1 (c), (d), for clarity, only two adjacent curves that form the grooves on the fixed member for each transmission scheme are shown. As shown in figure 1 (c), (d), each of the two curves is connected at one end to the curve forming the groove on the drive shaft. In this way, the grooves on the drive shaft and the fixed shaft are closed. The driven shaft is a part where intermediate rolling bodies (composite rollers) are fixed. Composite rollers are cylindrical disks with protrusions that interact with the grooves on the drive shaft and the fixed member. The minimum number of protrusions is two; one protrusion must be in contact with the groove on the drive shaft, and the other must be in contact with the groove on the fixed member.

The maximum possible number of composite rollers is determined according to the relationship:

$$n_{max} = Z_3 / Z_1 + Z_1 \tag{1}$$

where Z_3/Z_1 must be a whole number.

If necessary, a smaller number of composite rollers can be used in the transmission; this number should correspond to any divider for the number n_{max}.

The rotation of the drive shaft makes the composite roller to rotate due to interaction between one of its protrusions and the groove on the drive shaft. In this case, the other protrusion of the rotating composite roller interacts with the groove on the fixed member. As a result, the driven shaft, in which the composite rollers are installed, rotates at a lower speed. During the rotation of the composite roller, each protrusion moves successively from the groove on the drive shaft to the groove on the fixed member, i.e., it moves along the closed groove consisting of the helix (Archimedean spiral) on the drive shaft and the groove on the fixed member.

The reduction ratio is calculated according to the formula:

$$U = Z_1 \cdot Z_3 + 1 \tag{2}$$

The reduction ratio varies from 4 to 121. In this case, the transmission can have radial dimensions less than 200 mm while maintaining high load capacity.

3. Equations of helices used for cutting grooves

To cut grooves for both configurations of the transmission, equations of helices for a torus of the transmission parts have been derived.

The helix equations for the drive shaft of the cylindrical configuration of the transmission are as follows:

$$z=\frac{A_v}{2\cdot\pi}\cdot Z_1\cdot\theta_1 \tag{3}$$

$$x=\left\{R_{p1}+\left(r_s-\left[r_s^2-(r_s-z)^2\right]^{1/2}\right)\right\}\cdot\cos(Z_1\cdot\theta_1) \tag{4}$$

$$y=\left\{R_{p1}+\left(r_s-\left[r_s^2-(r_s-z)^2\right]^{1/2}\right)\right\}\cdot\sin(Z_1\cdot\theta_1) \tag{5}$$

where z, x, y are the coordinates of the points of the helix, m; A_v is the pitch of the helix, m; θ_1 is the central angle (varies in the range from 0 to 2π), rad; R_{p1} is the minimum distance from the axis of rotation to the torus, m; r_s is the radius of the circle used to obtain the torus, m (see figure 1 (c).

The helix equations for the fixed member of the cylindrical configuration of the transmission are as follows:

$$z=\frac{r_s\cdot Z_3}{\pi}\cdot\theta_3 \tag{6}$$

$$x=\left\{R_{p3}-\left(r_s-\left[r_s^2-(r_s-z)^2\right]^{1/2}\right)\right\}\cdot\cos\theta_3 \tag{7}$$

$$y=\left\{R_{p3}-\left(r_s-\left[r_s^2-(r_s-z)^2\right]^{1/2}\right)\right\}\cdot\sin\theta_3 \tag{8}$$

where θ_3 is the central angle (varies in the range from 0 to $2\pi/Z_3$), rad; R_{p3} is the maximum distance from the axis of rotation to the torus, m (see figure 1 (c).

The helix equations for the drive shaft of the radial configuration of the transmission are as follows:

$$A_v=\frac{R_{p3}-a}{Z_1\cdot 2\pi} \tag{9}$$

$$z=\left\{\left(\frac{R_{p3}-a}{2}\right)^2-\left\{\left(\frac{R_{p3}-a}{2}\right)-\left[(a+A_v\cdot Z_1\cdot\theta_1)-\left(\frac{R_{p3}-a}{Z_1}\right)\right]\right\}^2\right\}^{1/2} \tag{10}$$

$$x=(a+A_v\cdot Z_1\cdot\theta_1)\cdot\cos(Z_1\cdot\theta_1) \tag{11}$$

$$y=(a+A_v\cdot Z_1\cdot\theta_1)\cdot\sin(Z_1\cdot\theta_1) \tag{12}$$

where a is the initial radius of the Archimedean spiral, m (see figure 1 (d).

The helix equations for the fixed member of the radial configuration of the transmission are as follows:

$$z=-\left\{\left(\frac{R_{p3}-a}{2}\right)^2-\left\{\left(\frac{R_{p3}-a}{2}\right)-\left[\left(a+\frac{Z_3\cdot(R_{p3}-a)}{2\pi}\cdot\theta_3\right)-\left(\frac{R_{p3}-a}{Z_1}\right)\right]\right\}^2\right\}^{1/2} \tag{13}$$

$$x=\left\{a+\frac{Z_3\cdot\left[R_{p3}-a\right]}{2\pi}\cdot\theta_3\right\}\cdot\cos\theta_3 \tag{14}$$

$$y=\left\{a+\frac{Z_3\cdot\left[R_{p3}-a\right]}{2\pi}\cdot\theta_3\right\}\cdot\sin\theta_3 \tag{15}$$

4. Substantiation of the rational design of the transmission

3D models of cylindrical and radial configurations of the planetary torus helical transmission and the composite rollers are presented in figure 2.

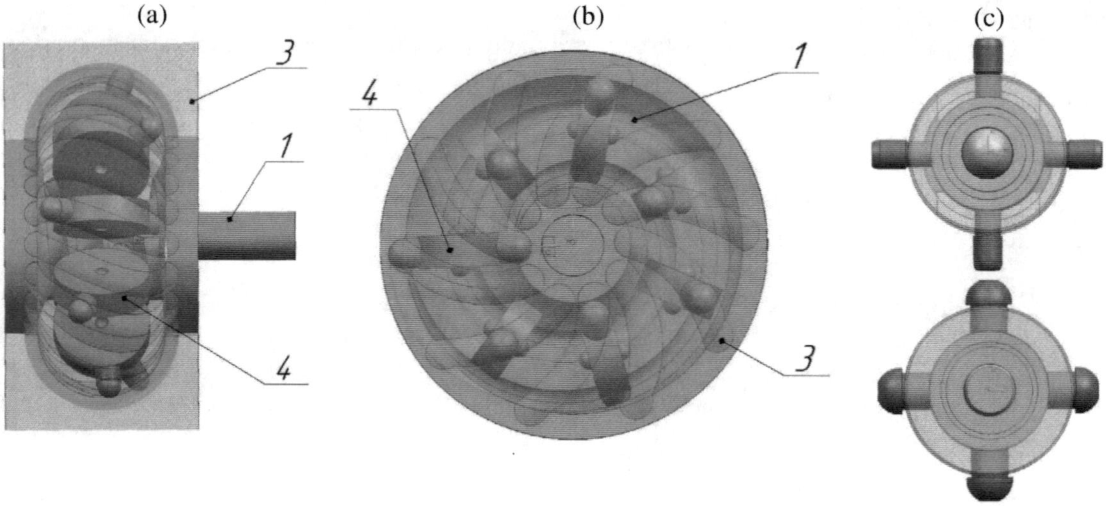

Figure 2. 3D models of cylindrical (a) and radial (b) configurations of planetary torus helical transmission (driven shaft not shown) and composite rollers (c): 1 – drive shaft; 3 – fixed member; 4 – composite rollers.

The comparison of the designs of the transmissions with cylindrical and radial configurations shows that, with equal dimensions and efficiency, the complexity of manufacturing and assembly (of the fixed member, in particular) is much higher for the cylindrical configuration than for the radial configuration. Moreover, in the case of the cylindrical configuration, the load capacity of the transmission is reduced because of the limited strength of the driven shaft due to cantilever arrangement of the protrusions where the composite rollers are fixed. Subsequently, the radial configuration of the planetary torus helical transmission will be studied as it is considered more rational.

The design of composite rollers has a significant impact on the power loss in the transmission [3] and its load capacity. The minimum number of the protrusions should be equal to two, but their number can be increased in order to increase the load capacity of the transmission, since the transmitted force is distributed proportionally to the number of the protrusions. The maximum possible number of the protrusions in the composite roller is $2 \cdot Z_1$. Hence, the number of the protrusions should be in the range from 2 to $2 \cdot Z_1$ and be divisible by 2. However, if the number of the protrusions is equal to $2 \cdot k$, the number of the grooves on the fixed member must also be increased k times. This increase does not affect the reduction ratio and the helix equations, but leads to a decrease in dimensions of the grooves and, consequently, the protrusions, which in turn affects their strength.

The protrusions can be cylindrical and spherical in shape (see figure 2 (c). From the point of view of manufacturability (grooves are cut with a spherical milling cutter), the spherical shape of the protrusions is more rational. However, at the same time a significant decrease in efficiency is observed (by 30 ... 50 %). Grooves for cylindrical protrusions can be cut with a spherical milling cutter by making multiple passes; in this way, a cylindrical surface is gradually formed. To obtain coordinates of the appropriate trajectories of the milling cutter in equations (9) – (15), the value R_{p3} is gradually increased and the value a is decreased in the same way. Obviously, this manufacturing method is more labor intensive.

5. Kinematic analysis of the transmission

To study kinematic parameters of the radial configuration of the planetary torus helical transmission, an NX CAD model of the transmission with the following parameters was created: $Z_1 = 2$; $Z_3 = 10$; $n = 7$; $R_{p1} = 0.042$ m; $a = 0.014$ m; the number of the composite roller protrusions is 2.

To simplify the calculations of rotation of the transmission parts, the fixed member was "released" and acted as a driven shaft, and the driven shaft was "fixed". In this case, the reduction ratio was calculated according to the formula: $U = Z_1 \cdot Z_3 = 20$. The drive shaft is rotating at a speed of 40 min^{-1}.

The graphs of angular velocities of rotation of the drive shaft, the driven shafts and the composite roller are presented in figure 3.

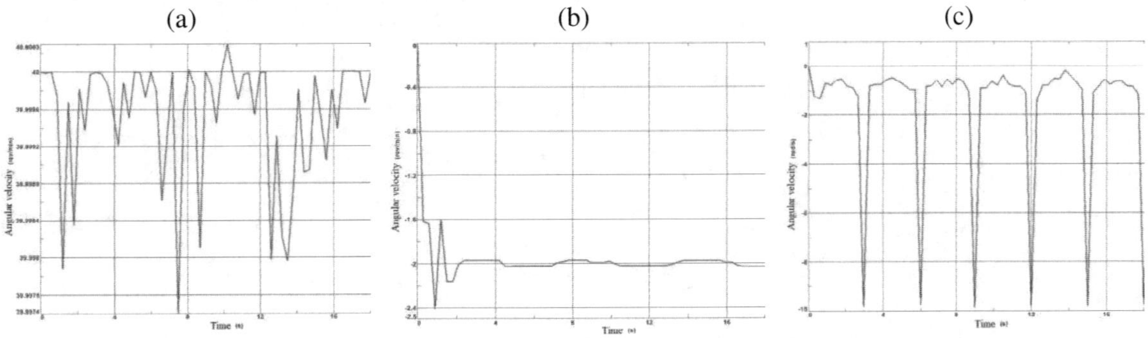

Figure 3. Graphs of angular velocities of drive shaft (a), driven shaft (b) and composite roller (c).

It has been found from the graphs in figure 3 (a) and (b) that the transmission has the required reduction ratio.

On the graph of the angular velocity of the composite roller, significant changes in the angular velocity are observed every 3 seconds, which corresponds to the point in time when the protrusion of the roller moves from the groove on one part to the groove on the other part of the transmission. These abrupt changes will lead to dynamic loads and noise during operation of the transmission.

6. Adjusting helix equations used for cutting grooves

To reduce abrupt changes in the angular velocity of the composite roller, it is necessary to adjust the helix equations for the drive shaft and the fixed member by making transition from one groove to another smoother [4].

The helix equations for the drive shaft of the radial configuration of the transmission are as follows:

$$z=\left\{\left(\frac{R_{p3}-a}{2}\right)^2-\left\{\left(\frac{R_{p3}-a}{2}\right)-\left(\frac{R_{p3}-a}{2}\right)\cdot\left[1-\cos\left(\frac{\theta_1}{2}\right)\right]\right\}^2\right\}^{1/2} \tag{16}$$

$$x=\left\{\left(\frac{R_{p3}-a}{Z_1}\right)+\left(\frac{R_{p3}-a}{2}\right)\cdot\left[1-\cos\left(\frac{\theta_1}{2}\right)\right]\right\}\cdot\cos\left(Z_1\cdot\theta_1\right) \tag{17}$$

$$y=\left\{\left(\frac{R_{p3}-a}{Z_1}\right)+\left(\frac{R_{p3}-a}{2}\right)\cdot\left[1-\cos\left(\frac{\theta_1}{2}\right)\right]\right\}\cdot\sin\left(Z_1\cdot\theta_1\right) \tag{18}$$

The helix equations for the fixed member of the radial configuration of the transmission are as follows:

$$z=-\left\{\left(\frac{R_{p3}-a}{2}\right)^2-\left\{\left(\frac{R_{p3}-a}{2}\right)-\left(\frac{R_{p3}-a}{2}\right)\cdot\left[1-\cos\left(\frac{Z_3\cdot\theta_3}{2}\right)\right]\right\}^2\right\}^{1/2} \tag{19}$$

$$x=\left\{\left(\frac{R_{p3}-a}{2}\right)\cdot\left[1-\cos\left(\frac{Z_3\cdot\theta_3}{2}\right)\right]+\left(\frac{R_{p3}-a}{Z_1}\right)\right\}\cdot\cos\theta_3 \tag{20}$$

$$y=\left\{\left(\frac{R_{p3}-a}{2}\right)\cdot\left[1-\cos\left(\frac{Z_3\cdot\theta_3}{2}\right)\right]+\left(\frac{R_{p3}-a}{Z_1}\right)\right\}\cdot\sin\theta_3 \qquad (21)$$

The graph of the angular velocity of the rotation of the composite roller after using the adjusted helix equations used for cutting grooves is shown in figure 4.

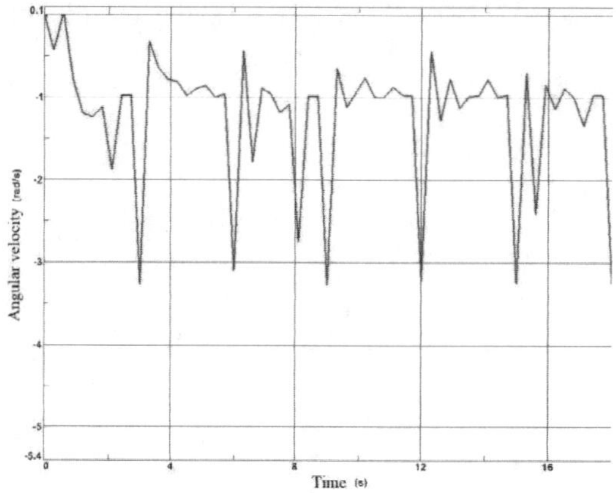

Figure 4. Graph of angular velocity of composite roller after adjusting helix equations

As a result, the magnitude of abrupt changes in angular velocity has decreased by a factor of 3.

7. Conclusion

The substantiation of the design of the planetary torus helical transmission providing high efficiency and load capacity with a reduction ratio in the range from 4 to 121 and a radial dimension of the transmission of not more than 200 mm has been performed. Equations of helices for obtaining grooves for the drive shaft and the fixed member that reduce dynamic loads and noise during operation of the transmission have been derived.

References

[1] Nam W K, Shin J W and Oh S H 2013 Design of thin plate-type speed reducers using balls for robots *Journal of Mechanical Science and Technology* **27**(2) pp 519–524

[2] Lustenkov M E 2010 Planetary ball transmissions: strength calculations *Russian Engineering Research* vol. 30 **9** pp 862–866

[3] Prudnikov A P 2018 Thermal analysis of transmission with intermediate rolling bodies *AER-Advances in Engineering Research (AviaENT 2018)* vol. 158 pp 338–342

[4] Efremenkov E A and An I-Kan 2010 Euler–Savari determination of radii of curvature of cycloid profiles *Russian Engineering Research* vol. 30 **10** pp 1001–1004

Influence of Wear-Resistant Coating on the Reliability of Replaceable Inserts in Cutting Steel

J Zhang, V N Kozlov and S Liu

National Research Tomsk Polytechnic University, 30, Lenina ave., Tomsk, 634050, Russia

E-mail: 965075948@qq.com

Abstract. Results of cutting tool life research of replaceable inserts (replaceable polyhedral plates) with wear-resistant coating and stressed-deformed condition of their cutting wedge are presented. Investigations have been executed in steel (0.40 % C, 1 % Cr) machining. Components of cutting force were measured by means of the dynamometer Kistler, magnitude and distribution of contact stresses have been defined according to earlier executed research in this area with use of a method of a sectional (split) cutting tool. Calculation of stresses in a cutting insert has been executed with use of program ANSYS which has displayed reduction of internal stresses in a base material of the insert with an antiwear coating in comparison with the insert without it. In the antiwear coating equivalent stresses more than in the insert without a coating and rich 1821 MPa. Experimental data on influence of various kinds of coats on cutting tool life and strength of replaceable inserts are presented.

1. Introduction

One of methods of working ability increasing of cutting tools is to use of wear-resistant coatings. Application of replaceable polyhedral cutting plates or replaceable inserts has allowed to reduce the cost price of deposition of high-quality coatings in a vacuum chamber and to improve their quality. To increase in reliability of work and durability of coatings they are applied multilayered [1-11].

Each layer of such composite coatings has own functionality. Properties of a composite cutting material with a coating can be steered due to a variation of a chemical compound of coatings, their structure and type of link with a base cutting material.

2. Materials and condition of experiments execution

Influence of cutting speed on tool life of inserts with three types of coating TiN, AlTiN, TiCN and without a coat was examined in turning steel (0.40 % C, 1 % Cr). The wear chamfer on a flank surface with the greatest its maximum permissible length $h_f = 0.2$ mm was accepted as criterion of wear of a cutting insert, corresponding for semifinish turning.

The cutting mode also corresponded to semifinish turning: feed rate $f = 0.13$ mm/r, depth of cut $t = 1$ mm, cutting speed $v = 360$ m/min. At execution of researches the change of a chamfer length h_f on a flank surface during of cutting (figure 1) was investigated.

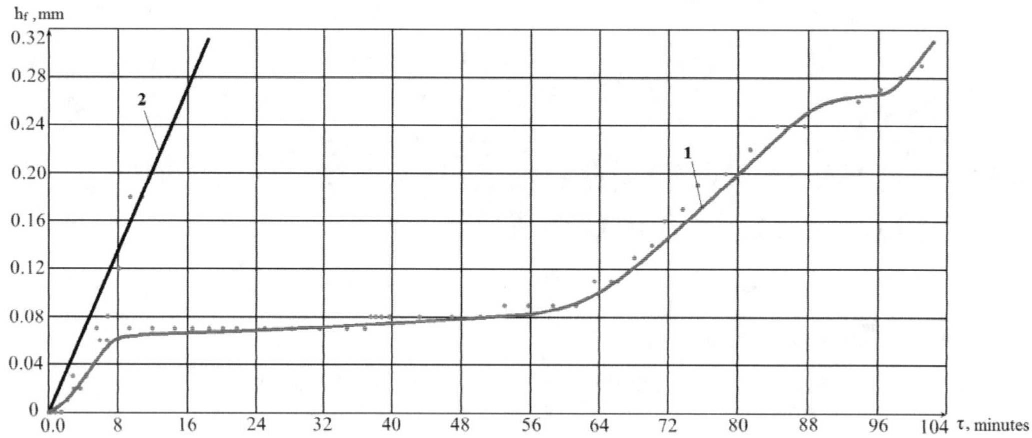

Figure 1. Dependence of length of the wear chamfer on the flank surface h_f (mm) from a time of cutting τ (minutes) in steel (0.40 % C, 1 % Cr) machining.

Steel (0.40 % C, 1 % Cr) – insert WNMG080404-TM with the coating TiN. f = 0.13 mm/r; t = 1 mm. 1 – v = 180 m/min; 2 – v = 360 m/min.

3. Research of wear

Character of change of a chamfer length during of experiments depends on cutting speed: at cutting speed 360 m/min it is rectilinear dependence on a time of cutting τ (min), but at cutting speed twice smaller (180 m/min) it is more complex dependence (figure 1). At cutting speed 180 m/min during 8 minutes the length of wear chamfer h_f is increased to 0.06 mm, after that the intensity of wear is sharply diminished that is connected, in our opinion, with the finish of wear-in time.

After 56 minutes of cutting the intensity of wear is sharply increased and becomes almost the same, as during the wear-in time.

After reaching of wear chamfer length h_f = 0.3 mm the chip is badly broken, is got confused round the workpiece and a cutting tool, and, having got on cams of a turning chuck, is scattered every which way, therefore trials should be stopped because of danger of deriving of a trauma of the human controller and aggravation of observation of the cutting process. Therefore the maximum permissible length of the wear chamfer is limited by not due to increasing of a cutting force and an elastic deformation (strain) of technological system, but due to aggravation of the chip shape.

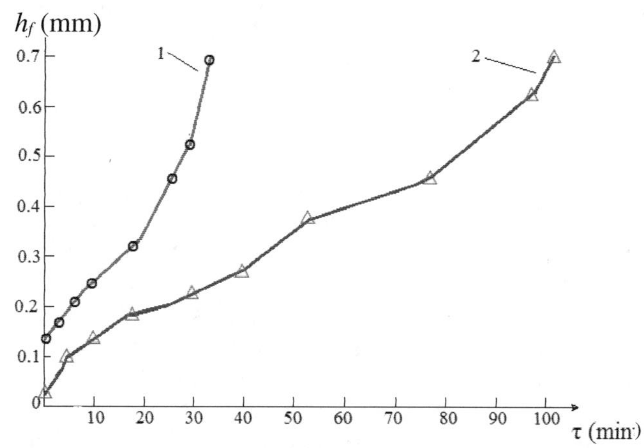

Figure 2. Change of length of the wear chamfer on the flank surface h_f (mm) of inserts made from cemented carbide (15%TiC, 6%Co, 79%WC) from a time of cutting τ (min) in steel (0.40 % C, 1 % Cr) machining. s = 0.13 mm/r; t = 1 mm; v = 300 m/min;
1 – inserts without a coat; 2 – inserts with the coat TiN.

Results of research of influence of machining time on the length of the wear chamfer h_f for inserts made from cemented carbide (15% TiC, 6% Co, 79% WC) with coat TiCN and without it are shown in figure 2. For inserts without coating at cutting speed $v = 300$ m/min the length of the wear chamfer $h_f = 0.3$ mm occur through 18 minutes. For inserts with coating TiCN the same wear occurs through 45 minutes that is 2.5 times longer.

More prolonged machining has allowed to augment the length of the wear chamfer h_f to 0.7 mm that corresponds to recommended maximum permissible wear for rough machining. Tool life without a coating and with the coating TiCN is 30 and 105 minutes accordingly, i.e. differ already more than 3 times.

In spite of the fact that after occurrence of the wear chamfer on the flank surface on this chamfer already there is no covering, all the same there is a protective action from a covering on a front surface which hangs over the chamfer a little and diminishes contact of transient surface with not already protected chamfer. This protective effect is more significant with the account of a sag of a transient surface in machining of the materials forming a continuous chip [1, 2, 3].

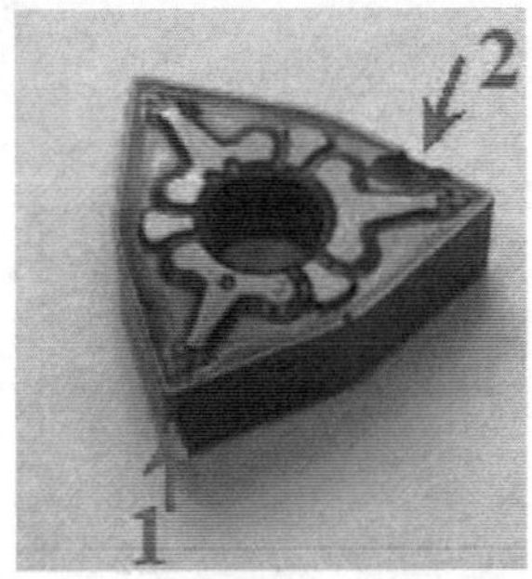

Figure 3. Wear of surface of the insert from a chip; 1 – working top of the insert, 2 – wear from the chip quenched on the air.

In machining with the cutting speed $v = 360$ m/min intensive local wear is observed on the surface which is positioned far from the working top (figure 3). It is explained by friction of the chip quenched after cooling on air.

Wear of a tool holder surfaces was not observed which is explained, in our opinion, by more essential reducing of temperature of the chip after cooling on a distance far from a cutting zone, thus smaller degree of effect.

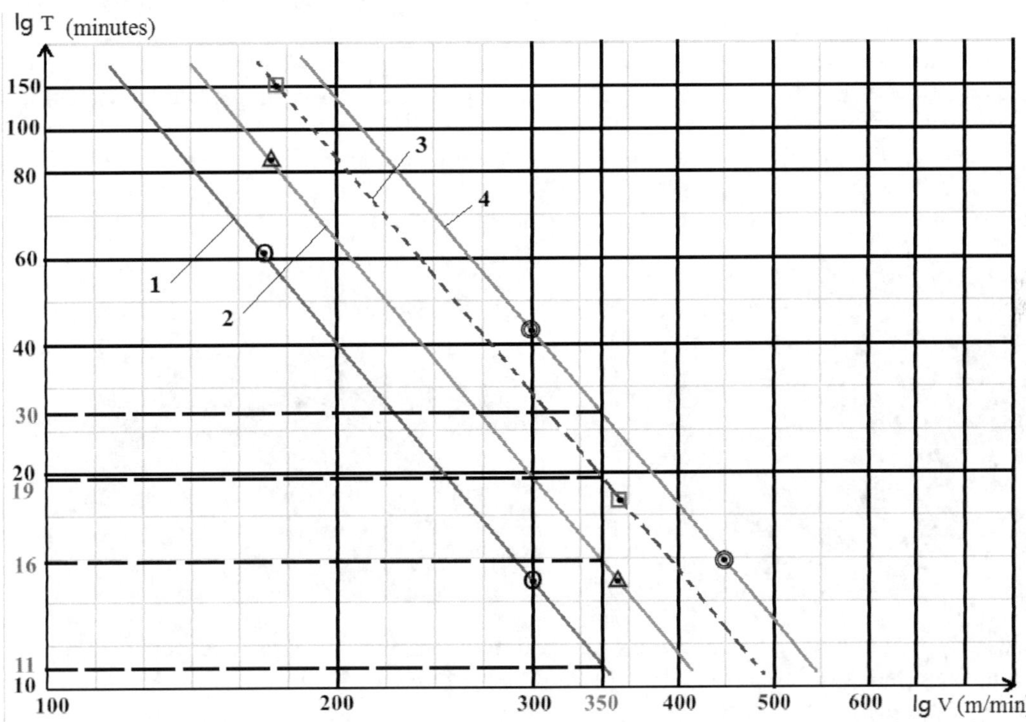

Figure 4. Dependence of the cutting tool life T (minutes) of inserts with wear-resistance coatings from cutting speed v (m/min) in steel (0.40 % C, 1 % Cr) machining and at h_f max = 0.2 mm: 1 – without a coat, 2 – with the coat TiN, 3 – with the coat AlTiN, 4 – with the coat TiCN.

Intensive local wear is observed in a zone of contact of the principal cutting edge with a work (raw) surface in spite of the fact that the workpiece before the experiment has been machined. It confirms position about essential influence of air on chemical wear at high temperature.

Such local wear, so-called "moustache", leads to formation of confused continuous chip which is twisted (is being confused) around a shank of the cutting tool and the workpiece, therefore it is very much trauma-danger and difficult for removing. In a greater degree it is appeared at rather small rate of cutting v = 180 m/min.

The maximum permissible length of the wear chamfer on the flank surface h_f = 0.2 mm has been selected for construction of tool life graphs $T = f(v)$ because at cutting speed v = 180 m/min the action from "moustache" leads to aggravation of observation of the cutting process. The experiments have displayed that the greatest tool life have inserts with covering TiCN (figure 4).

Tool life comparisons were executed also at cutting speed 360 m/min which corresponds to usually recommended cutting speed in semifinish machining, and at 180 m/min (corresponds to usually recommended cutting speed in rough machining with inserts).

4. Research of stresses in inserts

Loss of form of the cutting wedge may occur due to wear of the wedge on the face and flank, and also due to destruction: brittle destruction or plastic deformation. For definition of probability of destruction of cutting material and the covering the 3-D sample pieces (the model) of the insert have been created with the covering and without it (figure 4). External loads were set by the application of the contact stresses received from earlier executed experiments [1, 2, 3].

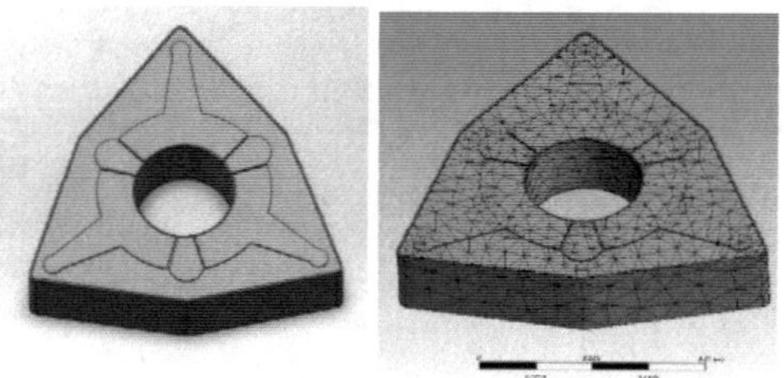

Figure 5. An example of the 3-D sample pieces of the cutting plate for calculation with a finite element method for the stressed-deformed assaying.

The stressed-deformed condition (SDC) of a cutting wedge has been analyzed by a finite element method (FEM) with use of program ANSYS. Distribution of internal equivalent stresses in the cutting insert is presented in figures 6 and 7.

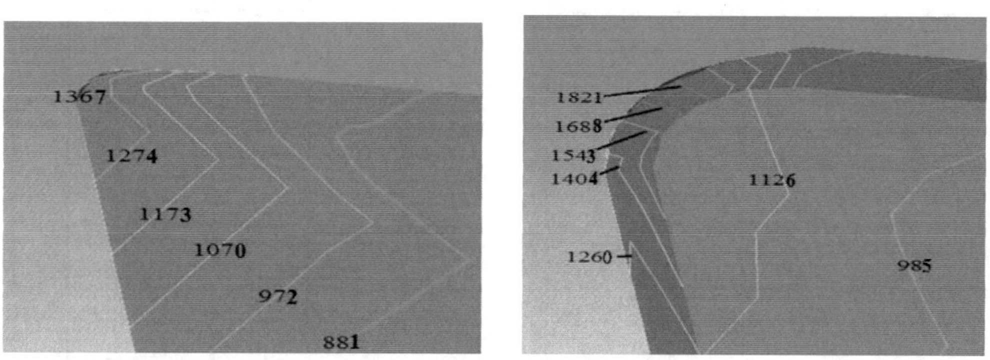

Figure 6. Distribution of equivalent stresses σ_e [MPa] in the cutting wedge of an acute cutting tool in turning steel. v = 180 m/min, f = 0,13 mm/r, t = 1 mm, h_f = 0 mm
a – the insert made from cemented carbide T15K6 without covering; b – the insert made from cemented carbide T15K6 with covering TiN.

The SDC analysis has displayed that the cutting plate with the coating has smaller internal equivalent stresses ($\sigma_{e\ max\ wedge\ with\ coat}$ = 1126 MPa) in comparison with the cutting plate without coating ($\sigma_{e\ max\ wedge\ without\ coat}$ = 1367 MPa), though in the coat layer equivalent stresses more than in a plate without a covering ($\sigma_{e\ max\ in\ coat}$ = 1821 MPa against $\sigma_{e\ max\ wedge\ without\ coat}$ = 1367 MPa). This result is explained by us due to increased Young coefficient of elasticity E of the coat TiN. In the coat at its depositing by the method PVD and consequently its heightened strength due to small amount of imperfections even at its small depth allow to withstand even big stresses.

At length of wear chamfer $h_f = 0.11$ mm the plate with the coat has 1.4 times more the greatest stress ($\sigma_{e\ max\ wedge\ with}$ coat = 2508 MPa) in comparison with the insert without wear on the flank surface ($h_f = 0$ mm) ($\sigma_{e\ max\ wedge\ with\ coat} = 1821$ MPa).

However at rather large wear ($h_f = 0.2$ mm) the insert with the coat has the greatest internal stress ($\sigma_{e\ max\ coat} = 1888$ MPa) close to the insert without wear ($h_f = 0$ mm) ($\sigma_{e\ max\ wedge\ with\ coat} = 1821$MPa). Therefore the antiwear coating executes protective function on the flank surface even at rather large wear $h_f = 0.2$ mm.

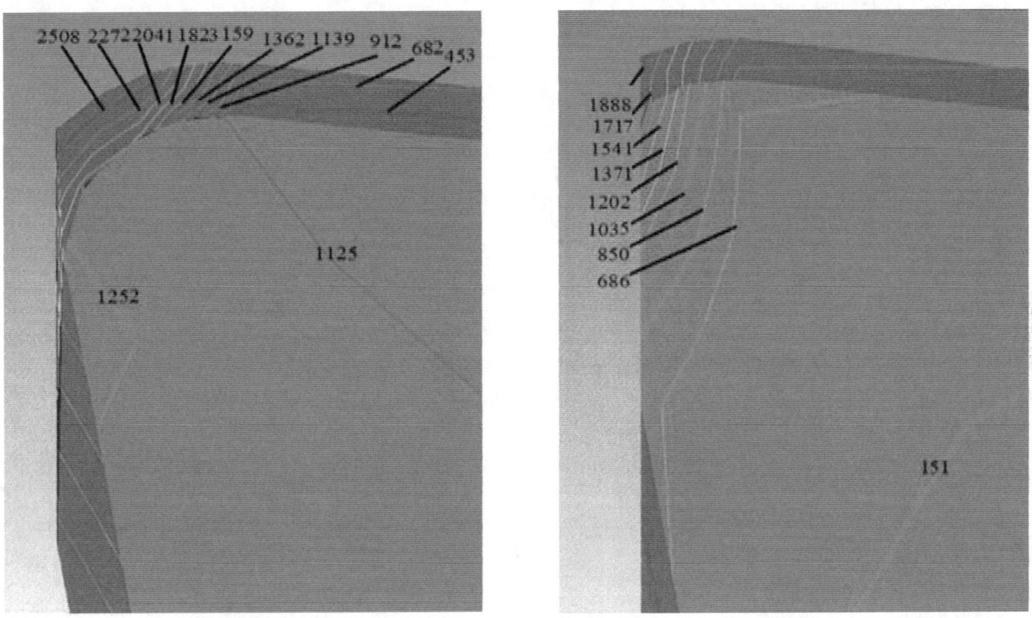

a b

Figure 7. Distribution of equivalent stresses σ_e [MPa] in the cutting wedge of the insert made from cemented carbide T15K6 with covering TiN in turning steel. $v = 180$ m/min, $f = 0{,}13$ mm/r, $t = 1$ mm. a – $h_f = 0.11$ mm; b – $h_f = 0.2$ mm.

5. Conclusions

1. Character of change of a chamfer length during of experiments depends on cutting speed: at cutting speed 360 m/min it is rectilinear dependence on a time of cutting, but at cutting speed twice smaller (180 m/min) it is more complex dependence.
2. At turning steel with cutting speed 300 m/min and feed rate 0.13 mm/r tool life of inserts made from cemented carbide with the coat TiCN exceeds tool life of inserts without the coat 2.5-2.9 times;
3. Internal stresses in a cutting wedge of the insert with antiwear coat TiN are diminished by 18 % (on 18 %) in comparison with the insert without a coat;
4. At turning steels with cutting speed $v = 350$ m/min, s = 0.13 mm/r; t = 1 mm cutting tool life of cemented carbide inserts with the coat TiCN exceeds control inserts without a coat 3 times.
5. The greatest internal stress in the coat on 25 % more than in the cemented carbide insert without a coat that is connected with small depth of the coat;

6. Inserts with the coat TiCN has minimal floor price of machining of 1 metre. From economic feasibility point of view application of inserts with the coat TiN it is better, than with the coat AlTiN. Recommendations for choice an antiwear coating for semifinish machining of a steel: inserts with the coat TiCN and inserts with the coat TiN.

Acknowledgments
The research is carried out at National Research Tomsk Polytechnic University within the framework of National Research Tomsk Polytechnic University Competitiveness Enhancement Program grant.

References
[1] Kozlov V, Zhang J, Cui J and Bogolyubova M 2017 *Key Engineering Materials. Trans Tech Publications* **73** 252-257
[2] Kozlov V, Zhang J, Guo Y and Sabavath S K 2018 *High Technology: Research and Applications* **769** 284-289
[3] Kozlov V N 2012 *Proc. 7th Int. Forum on Strategic Technology* **2** 147-151
[4] Artamonov E V, Chernyshov M O. and Pomigalova T E 2017 *Russian Engineering Research* **37**, No. 4 348-350
[5] Hu J and Chou H K 2007 *Wear* **263**, Iss 7-12 SPEC. ISSS 1454-1458
[6] Merchant M E 2007 *J. Appl. Phys.* **16** Iss 5 267-275
[7] Trent E M and Wright P K 2000 *Metal Cutting* (Boston: Butterworth-Heinemann)
[8] Boothroyd G and Knight W 2006 *Fundamentals of Machining and Machine Tools*
[9] Proskokov A V and Petrushin S I 2012 *Proc. 7th Int. Forum on Strategic Technology* 173-177
[10] Afonasov A and Lasukov A 2014 *Russian Engineering Research* **3** 152-155
[11] Lasukov A A, Chazov P A and Barsuk A V 2014 *Applied Mechanics and Materials* **682** 504-509

IOP Conf. Series: Materials Science and Engineering **795** (2020) 012024 doi:10.1088/1757-899X/795/1/012024

Determination of Geometric Parameter of Cycloidal Transmission from Contact Strength Condition for Design of Heavy Loading Mechanisms

E A Efremenkov[1,2], S K Efremenkova[1] and I M Dyussebayev[3]

[1]National Research Tomsk Polytechnic University, 30, Lenina ave., Tomsk, 634050, Russia
[2]Tomsk State University of Control Systems and Radioelectronics, Tomsk, Lenin ave., 40, 634050, Russia
[3]Satbayev University, Almaty, Satbayev Street, 22, 050000, Kazakhstan

E-mail: ephrea@mail.ru

Abstract. In the article, the contact strength condition is considered for transmission with intermediate rolling bodies. The components of contact strength condition equation are expressed through initial parameters of the transmission. The expression is obtained for determination the minimum permissible value of the geometric parameter r_c of transmission with IRBFIR. The expression can be used in preliminary engineering calculations of the hard loaded transmission.

1. Introduction

In modern machines and mechanisms, transmissions with intermediate rolling bodies (IRB) are increasingly in demand. The transmissions are used in the oil and gas industry, thermal power plants, transport systems [1], mining [2]. In addition, IRB transmissions are used in drilling mud agitator drives, mining harvesters and other heavy loaded mechanisms. For highly loaded machines, the most important parameter is the high load capacity of the gears included in them and the ability to hold significant overloads while operability remain unchanged. Transmission with intermediate rolling bodies and free iron ring (IRBFIR) allows such characteristics to be provided to the mining machine.

Transmission with IRBFIR (figure 1) has complex of high technical characteristics provided by multi-pair engagement and sliding friction is reduced in engagement. In the design of mechanical power transmission, the cross section area of the transmission loading parts is the defining measure of strength, including for mining machines. And the cross section dimensions are determined based on the contact strength condition and affect the transmission dimensions itself and vice versa. Therefore, the determination of the transmission with IRBFIR geometric characteristic from the contact strength condition is relevant for the design of heavy loading mechanisms for mining operations and engineering calculations.

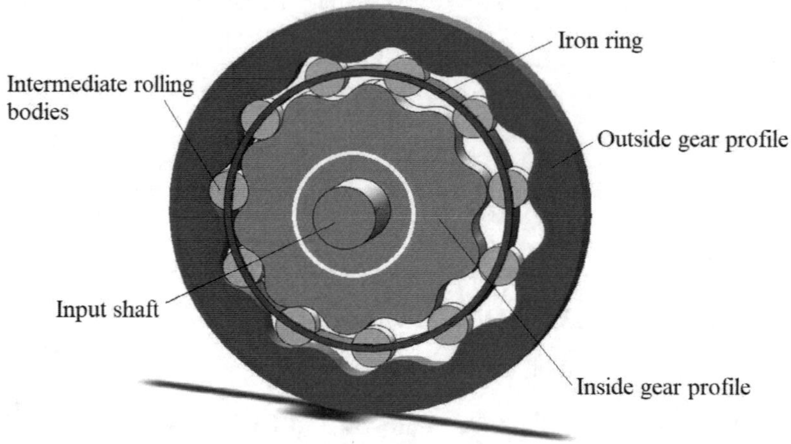

Figure 1. The scheme of transmission with intermediates rolling elements and free iron ring.

Transmissions with IRB have been studied for quite a long time [1, 3], but so far their using is limited. One reason is the lack of the gear sizes valid choice based on the contact strength condition. Scientists in Russia, Belarus [3-7] are engaged in research of transmissions with IRB and calculation of contact stresses. Beyond the borders of the Soviet Union basic information about cycloidal gearing, geometry and force calculation procedure was presented by Lehmann [8], which later was extended in [9-12]. Blanche and Yang [13] investigated the influence of machining tolerance on drive performance indexes and their relationships with drive parameters. Dynamic behavior of cycloidal parts is considered in [14-16]. However, expression was not presented for determination of the transmission with IRBFIR geometric parameters based on the contact strength condition. Therefore, the purpose of the work is to obtain such an expression for determining the radius of the centres of the rolling bodies of the transmission with IRBFIR through the initial parameters. Initial parameters for the transmission are: r_2 – the radius of the making circle, Z_2 - number of bodies of swing, χ – shift factor and r_b –the radius of a rolling body [3].

2. Considering contact strength condition for transmission with IRBFIR

The contact strength condition is written as:

$$(\sigma_H)_{max} \le [\sigma_H],$$

where $(\sigma_H)_P$ – maximum design contact stress;

$[\sigma_H]$ – permissible contact stress.

Let us consider the equation for determining the contact stresses on the i-th rolling body of the transmission with IRBFIR [3]:

$$(\sigma_H)_i = \left(\frac{F_i(\rho_2 \pm \rho_{1i})}{\pi \cdot l_b \cdot \rho_{1i}\rho_{2i}\left(\frac{1-\mu_1^2}{E_1}+\frac{1-\mu_2^2}{E_2}\right)} \right)^{1/2}, \quad (1)$$

where F_i – normal force to surfaces of the toothed gear and i-th rolling body contact, H;

ρ_{1i}, ρ_2– curvature radii of contacting bodies (toothed gear and rolling body respectively), mm;

l_b – contact length of roller and profile, mm;

E_1, E_2 – elasticity modulus of first and second contacting bodies respectively, MPa;

μ_1, μ_2 – Poisson's ratios for first and second contacting bodies respectively.

Sign "+" is used for biconvex contact, but sign "-" is used for convexo-concave contact.

To determine the normal force, let us consider the force distribution diagram in engagement of transmission with IRBFIR (figure 2). Figure 2 presents: P – the pitch point; O_1, O_2, O_3 – centers: inside gear, iron ring with rolling bodies and outside gear, respectively; r_1, r_2, r_3 – the centroid radii of the inside gear, iron ring with rolling bodies and outside gear, respectively; r_c – radius of rolling body centers position; φ_2 –angle of the iron ring rotation (together with rolling bodies); e_1 - eccentricity of engagement; e - hollow eccentricity of the transmission; F_i - force in the transmission engagement on i-th rolling body; h_i – the shortest distance from the center of the inside gear to the line of action of the i-th force.

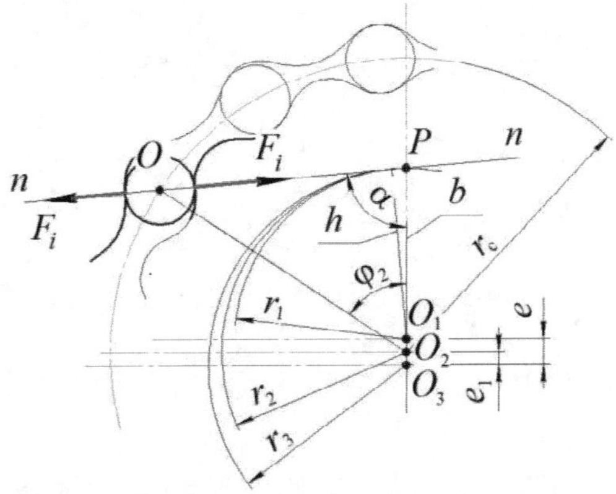

Figure 2. The calculation diagram for determination force into engagement of transmission with intermediates rolling elements and free iron ring

The torque on the inside gear of the transmission with IRBFIR is defined as

$$T_k = \sum F_i \cdot h_i.$$

It is known [8] that the maximum force $F_{\max}$, will be when α=90° (figure 2), then

$$F_{\max} = \frac{T_k \cdot b}{\sum h_i^2}$$

and

$$\frac{F_i}{h_i} = \frac{F_{\max}}{b} \rightarrow F_i = \frac{F_{\max} h_i}{b}.$$

Therefore, for normal force in contact with the i-th rolling body the expression through torque is recorded as

$$F_i = \frac{T_k \cdot b}{\sum h_i^2} \cdot \frac{h_i}{b} = \frac{T_k h_i}{\sum h_i^2}. \quad (2)$$

Assuming that the cycloidal gears and rolling bodies are made of the same material, as is generally the case. Therefore, the contact strength condition taking into account the expressions (1) and (2) will be recorded as:

$$(\sigma_H)_{\max} = \left(\frac{T_k h_i E(\rho_{2i} \pm \rho_{1i})}{2\pi \cdot l_b \cdot \rho_{1i} \rho_{2i} (1-\mu^2) \sum h_i^2}\right)^{1/2} \le [\sigma_H]. \quad (3)$$

3. Determination of the components of contact strength condition equation

Radii of curvature are determined in [9] through initial parameters of transmission with IRBFIR. Thus, the radius of curvature of the cycloidal profile of the inside gear is expressed through initial parameters as

$$\rho_1 = r_2(1+\chi^2-2\chi\cos\varphi)^{1/2} - r_b - Z_2 r_2 i_{21}(1+\chi^2-2\chi\cos\varphi)^{1/2} \cdot \left(\chi Z_1 \cos\varphi + \frac{\chi^2 \cdot \sin^2\varphi}{(1-\chi\cos\varphi)} + Z_2(1-\chi\cos\varphi)\right)^{-1}, \quad (4)$$

where i_{21} is the gear ratio from the rolling bodies to the inside gear and it is defined from the expression:

$$i_{21} = 1 - \frac{1}{Z_2}.$$

The expressions of sum and product can be represented for the radii of curvature of the transmission with IRBFIR inside cycloidal gear and rolling body through the initial parameters as follows:

$$\rho_2 + \rho_1 = r_2(1+\chi^2 - 2\chi\cos\varphi)^{1/2} - Z_2 r_2 i_{21}(1+\chi^2-2\chi\cos\varphi)^{1/2} \cdot \left(\chi Z_1\cos\varphi + \frac{\chi^2\cdot\sin^2\varphi}{(1-\chi\cos\varphi)} + Z_2(1-\chi\cos\varphi)\right)^{-1};$$

$$\rho_2 \cdot \rho_1 = r_b r_2(1+\chi^2 - 2\chi\cos\varphi)^{1/2} - r_b^2 - r_b Z_2 r_2 i_{21}(1+\chi^2-2\chi\cos\varphi)^{1/2} \cdot \left(\chi Z_1\cos\varphi + \frac{\chi^2\cdot\sin^2\varphi}{(1-\chi\cos\varphi)} + Z_2(1-\chi\cos\varphi)\right)^{-1}.$$

For brevity we will designate $a = (1+\chi^2-2\chi\cos\varphi)^{1/2}$. Then after the transformations, we will get:

$$\rho_2 + \rho_1 = r_2 a\left(1 - Z_2 i_{21}\cdot\left(\chi Z_1\cos\varphi + \frac{\chi^2\cdot\sin^2\varphi}{(1-\chi\cos\varphi)} + Z_2(1-\chi\cos\varphi)\right)^{-1}\right); \quad (5)$$

$$\rho_2 \cdot \rho_1 = r_b r_2 a\left(1 - Z_2 i_{21}\cdot\left(\chi Z_1\cos\varphi + \frac{\chi^2\cdot\sin^2\varphi}{(1-\chi\cos\varphi)} + Z_2(1-\chi\cos\varphi)\right)^{-1} - \frac{r_b}{r_2 a}\right). \quad (6)$$

Denote here $k = 1 - Z_2 i_{21}\cdot\left(\chi Z_1\cos\varphi + \frac{\chi^2\cdot\sin^2\varphi}{(1-\chi\cos\varphi)} + Z_2(1-\chi\cos\varphi)\right)^{-1}$, then the expressions (5) and (6) are rewritten as

$$\rho_2 + \rho_1 = k r_2 a;$$

$$\rho_2 \cdot \rho_1 = r_b r_2 a\left(k - \frac{r_b}{r_2 a}\right).$$

The ratio of the sum of the radii of curvature to their product is written as

$$\frac{\rho_2+\rho_1}{\rho_2\cdot\rho_1} = \frac{k r_2 a}{r_b r_2 a}\cdot\left(k - \frac{r_b}{r_2 a}\right)^{-1}.$$

Let us finally receive

$$\frac{\rho_2+\rho_1}{\rho_2\cdot\rho_1} = \frac{k}{r_b}\cdot\left(k - \frac{r_b}{r_2 a}\right)^{-1}. \quad (7)$$

The expression of distance h is written through the initial parameters as:

$$h = \frac{i_{21} r_2 \chi\cdot\sin\varphi}{(1+\chi^2-2\chi\cos\varphi)^{1/2}}.$$

Then the ratio of distance (to the i-th normal) to the sum of squares of all distances is expressed as:

$$\frac{h_i}{\sum h_i^2} = \frac{i_{21} r_2\chi\cdot\sin\varphi}{a i_{21}^2 r_2^2 \chi^2}\cdot\left(\sum\left(\frac{\sin\varphi}{a}\right)^2\right)^{-1}.$$

After conversion we finally get

$$\frac{h_i}{\sum h_i^2} = \frac{\sin\varphi}{a r_2 \chi i_{21}}\cdot\left(\sum\left(\frac{\sin\varphi}{a}\right)^2\right)^{-1}. \quad (8)$$

By substituting the resulting expressions (6) and (7) in (2), we obtain:

$$(\sigma_H)_i = \sqrt{\frac{T_k E\cdot k\cdot\sin\varphi}{2\pi\cdot l_b(1-\mu^2) r_b a r_2 \chi i_{21}}\left(\left(k - \frac{r_b}{r_2 a}\right)\cdot\sum\left(\frac{\sin\varphi}{a}\right)^2\right)^{-1}}. \quad (9)$$

In expression (9) the product $r_2\chi$ determines the radius of the rolling bodies centers position r_c which is one of the geometrical indicators defining overall dimensions of the transmission. We believe that the gears and rolling bodies are made from steel. Then let us remove the constant values from the root and write the contact strength condition for transmission with IRBFIR as:

$$(\sigma_H)_{max} = 191{,}65 \cdot 10^3 \sqrt{\frac{T_k \cdot k \cdot \sin\varphi}{l_b r_b a r_c i_{21}} \left(\left(k - \frac{r_b}{r_2 a}\right) \cdot \Sigma \left(\frac{\sin\varphi}{a}\right)^2 \right)^{-1}} \leq [\sigma_H]. \quad (10)$$

From contact strength condition (10) we express radius of radius of the rolling bodies centers position r_c through permissible contact stress

$$r_c \geq \frac{36{,}73 \cdot 10^9 \cdot T_k \cdot k \cdot \sin\varphi}{l_b r_b a i_{21} [\sigma_H]^2} \left(\left(k - \frac{r_b}{r_2 a}\right) \cdot \Sigma \left(\frac{\sin\varphi}{a}\right)^2 \right)^{-1}. \quad (11)$$

4. Example

Let us calculate the radius of the rolling bodies centers position on the basis of the following initial parameters of transmission with IRBFIR: r_2=25 mm; Z_2=25; χ=1,4; r_b=3 mm. Let torque on the inside gear T_k=200Hm and l_b=6 mm, and permissible contact stress for steel ShH15 equally $[\sigma_H]$ =3000MPa.

In transmission with IRBFIR the maximum contact stress exists at angle $\varphi \approx 70°$. Therefore, all calculations will be made for this angle.

Then perform the calculation using the expressions (7), (8) and (11), obtain the following minimum allowable value of the radius of the rolling body centers position:

$$r_c \geq 0.032\, m.$$

Knowing that the radius r_c is connected to radius r_2 through shift factor χ [3], it is possible to specify centrode radius r_2

$$r_c = r_2 \cdot \chi \geq 0{,}032,$$

$$r_2 \geq \frac{0{,}032}{\chi},$$

$$r_2 \geq 0{,}023\, m.$$

Therefore, preselected initial parameters and, in particular the centroid radius r_2, satisfy the strength condition.

5. Conclusions

Thus, the expression (11) is obtained for determination the minimum permissible value of the geometric parameter r_c of transmission with IRBFIR. In the future, using this parameter it is possible to adjust initial parameters of the transmission taking into account strength. The expression obtained is basic for strength calculation of transmission with IRBFIR and mechanisms based on it and can be used in preliminary engineering calculations of the transmission.

References:

[1] Pankratov E N 1998 Designing of mechanical systems of the automated complexes for mechano-machining of manufacture. A practical work of the leader-designer *Tomsk State University Press* p 296

[2] Aksenov V V, Efremenkov A B, Blaschuk M Yu and Ryltseva Ya G 2012 *J. Vestnik Nauki Sibiri (Messenger of Siberia Science)* **1 (2)** 372-378

[3] Efremenkov E A 2002 Development of methods and means of increase of effectiveness of transmissions with intermediate rolling bodies: thesis on competition degree of candidate of technical science *Tomsk, TPU*

[4] An I-Kan, Il'in A S and Lazurkevich A V 2016 Load analysis of the planetary gear train with intermediate rollers. Part 2 *IOP Conf. Series: Materials Science and Engineering* **124**

[5] Ivkina O P, I-Kan A, Cheremnov A V and Pashkov E N 2012 Exposing static indeterminacy of dimensioned gear with packing rolling element *Tomsk, Russia, 7th International Forum on Strategic Technology* 6357717

[6] Lustenkov M E 2010 *J. Russian Eng. Res.* **30 (9)** 862-866

[7] Prudnikov A P 2018 *J. AER-Adv. in Eng. Res.* **158** 338–342

[8] Lehmann M 1976 Calculation and measurement of forces acting on cycloidal speed reducer *PhD Thesis* Technical University Munich, Germany
[9] Malhotra S K and Parameswaran M A 1983 Analysis of a cycloidal speed reducer *Mechanism and Machine Theory* **18 (6)** 491-499
[10] Yan H S and Lai T S 2002 Geometry design an elementary planetary gear train with cylindrical tooth profiles *Mechanism and Machine Theory* **37 (8)** 757–767
[11] Li X, He W, Li L and Schmidt L C 2004 A new cycloid drive with high-load capacity and high efficiency *ASME Journal of Mechanical Design* **126** 683–686
[12] Shin J H and Kwon S M 2006 On the lobe profile design in a cycloid reducer using instant velocity center *Mechanism and Machine Theory* **41** 596–616
[13] Blanche J G and Yang D C H 1989 Cycloid drives with machining tolerances *ASME Journal of Mechanisms Transmissions and Automation in Design* **111** 337–344
[14] Blagojević M, Nikolić V, Marjanović N and Veljović LJ 2009 Analysis of cycloid drive dynamic behavior *Scientific Technical Review* **LIX (1)** 52-56
[15] XiaoXiao Sun and Liang Han 2019 A New Numerical Force Analysis Method of CBR Reducer with Tooth Modification Speed Reducer *IOP Conf. Series: Journal of Physics: Conf. Series* **1187** 032053
[16] Li Xin Xu 2018 A dynamic model to predict the number of pins to transmit load in a cycloidal reducer with assembling clearance *Proc IMechE Part C: J Mechanical Engineering Science* 0 (0) 1–23
[17] Efremenkov E A 2009 Determination of forces in transmission with intermediate rolling bodies and free cage *6th Intersectoral Scientific and Technical Conf. "Automation and Advanced Technologies in the Nuclear Industry", Novouralsk NSTI* pp 123-126
[18] E.A. Efremenkov and An I-Kan 2010 *Russian Engineering Research.* **30 (10)** 1001-04

Power Parameters Automated Calculation for Transmission with Intermediate Rolling Bodies and Free Cage

E A Efremenkov[1,2] and E Bonnard[3]

[1]National Research Tomsk Polytechnic University, 30, Lenina ave., Tomsk, 634050, Russia
[2]Tomsk State University of Control Systems and Radioelectronics, Tomsk, Lenin ave., 40, 634050, Russia
[3]University Savoie Mont Blanc, Chambery, 27 Rue Marcoz, 73000, France

E-mail: ephrea@mail.ru

Abstract. In the article, techniques complex of the transmission with IRBFC power parameters determination are considered. Moreover, software of automated calculation of the power parameters is created based on the techniques. The software consists of two parts: program of the mechanical transmission creating as a computer object; program of the graphical user interface creating. In general, the software makes it possible to determine the distribution of forces and contact stresses in engagement along the cycloidal profile of gear, to determine the maximum force and contact stress in engagement and to generate a file in excel format with all calculated data.

1. Introduction

In modern industry, transmissions with intermediate rolling elements (IRE) are increasingly in demand. They are used in power plants, in pipelines valves of oil and gas systems, in transport drives, in aircraft mechanisms [1], in drives for mining industry [2]. These transmissions were studied by many authors [3-14]. Litvin and Feng [15] developed computer programs for generating the planar epicycloidal gearings. However, a transmission with intermediate rolling bodies and free cage (IRBFC) is the most promising for use in heavily loaded mechanisms. In connection with the growing need for the design of mechanisms based on transmission with IRBFC (figure 1), there is a need to accelerate the calculations and design of such transfers and mechanisms in general. For transmission with IRBFC operating in difficult conditions, the power parameters that this transmission is capable to provide are not unimportant. For cycloidal transmissions, contact stress is the main one. However, when performing calculations of efforts and contact stresses in the gear engagement "by hand", the probability of an error and the receipt of an incorrect result are high. Therefore, the automation of the calculation of power parameters of transmission with IRBFC and the development of the calculation program is relevant.

The transmission with IRBFC design starts from initial parameters. For the transmission initial parameters are: R_{21} – the radius of the making circle, Z_2 - number of bodies of swing, χ – shift factor and R_{RB} –the radius of a rolling body [16].

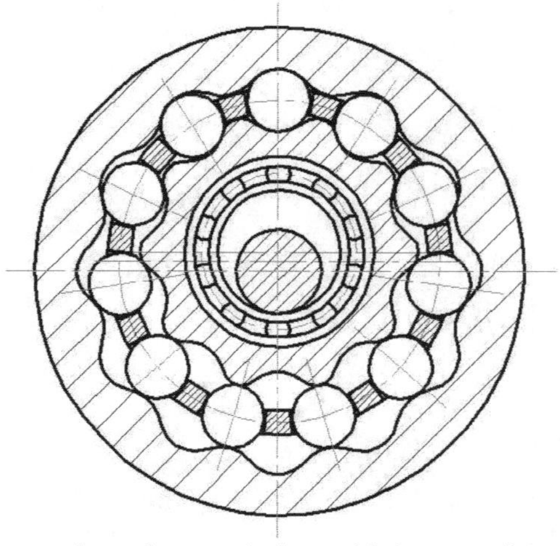

Figure 1. Cross section of transmission with intermediate rolling bodies.

2. Choosing equations

The determination of the forces in the transmission with IRBFC was carried out in [17], as:

$$F_i = \frac{T_c h_i}{\sum h_i^2}, \tag{1}$$

where T_c – torque of the cam, Hm;
h_i – the shortest distance from the center of the cam to the line of action of the i-th force.

The determination of contact stresses was described in [16]. And result expression we wright as

$$(\sigma_H)_i = \left(\frac{F_i(\rho_2 \pm \rho_{1i})}{\pi \cdot l \cdot \rho_{1i}\rho_{2i}\left(\frac{1-\mu_1^2}{E_1} + \frac{1-\mu_2^2}{E_2} \right)} \right)^{1/2}, \tag{2}$$

where ρ_1, ρ_2 – curvature radii of contacting bodies, mm;
l – contact length of roller and profile, mm;
E_1, E_2 – elasticity modulus of first and second contacting bodies respectively, MPa;
μ_1, μ_2 – Poisson's ratios for first and second contacting bodies respectively.

The sequence of calculating the radii of curvature of the profiles was decided to be taken from [18]. for cam wheel profile radius of curvature is

$$\rho_1 = l_{OP} - r_b - l_{MP}, \tag{3}$$

where l_{OP} – distance between of the rolling body center O and engagement pole P (figure 2), mm;
l_{MP} – distance between of the radius of curvature M and engagement pole P (figure 2), mm.

And for crown wheel profile radius of curvature is

$$\rho_3 = l_{MP} - r_b - l_{OP}, \tag{4}$$

Summarizing the methods presented in the sources we can obtain a program for automated calculation of loading parameters of the transmission with IRBFC. A program was developed for calculating the forces and contact stresses in the transmission engagement, the interface of which is shown in figure 1.

IOP Conf. Series: Materials Science and Engineering **795** (2020) 012025 doi:10.1088/1757-899X/795/1/012025

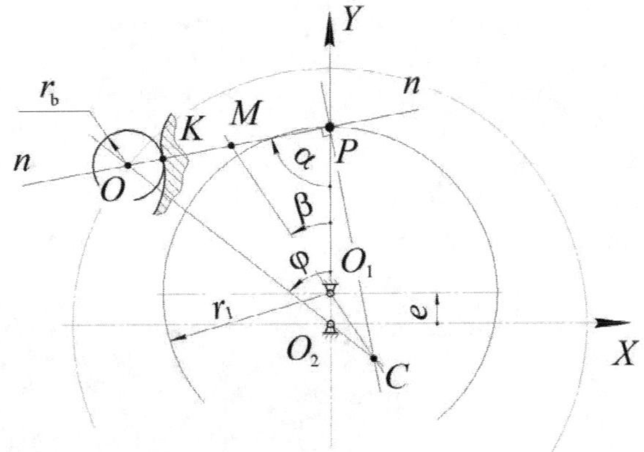

Figure 2. For cam profile to the determination of the radius of curvature.

3. Making an automated calculation program

The goal was to create a program that can be used on all types of computers with any operating system. This objective allowed choosing the computer language that would be used; in this case, the language was JAVA. On the other hand, this language requires the installation of java on the physical target (computer). This language is one of the most used in the computer world to create desktop applications because it is very powerful (belongs to object-oriented programming language). On Internet there are many resources and forums concerning this language such as "Stack overflow" or "Git Hub", which allows solving problems related to programming quickly.

To realize the computer object, the software used was "Eclipse" which is one of the best known to develop in JAVA. For the graphical interface, it was the software "NetBeans", which has a basic graphical interface designer.

The program allows performing calculation of contact strength of details loaded for transmission with IRBFC. The software allows us to change quickly the input parameters, if necessary. The program was developed by equations (1)-(4) for calculating the forces and contact stresses in the engagement of transmission with IRBFC, the interface of which is shown in figure 3.

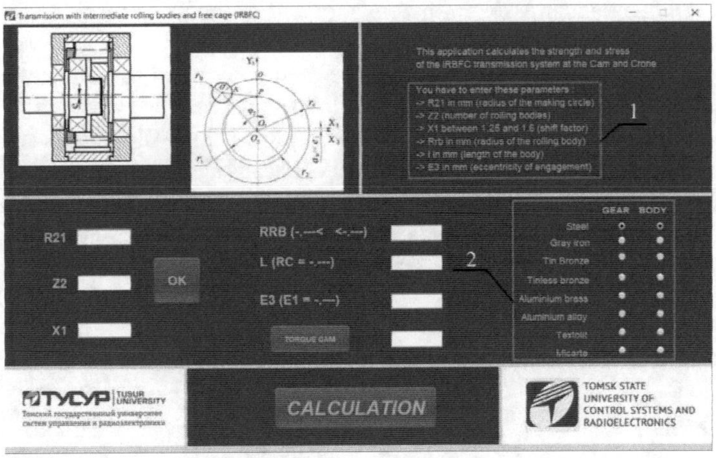

Figure 3. The interface of the program for the automated calculation of power transmission parameters with IRBFC.

The interface of the calculation program (figure 3) explains the meaning of the entered parameters and the dimension in which they should be entered into the program is indicated by frame 1. In addition, the program provides the ability to set the material of gear wheels and rolling elements of the gear, this is located in the lower right part of the interface (figure 3, frame 2).

In the process of working with the program, it is initially necessary to enter the values of three initial parameters: the radius of the generating circle R_{21}, the number of rolling bodies Z_2, and the shift factor χ. After "OK" button is pressed, intermediate parameters (radius of the rolling bodies' centers and eccentricity) and range of possible values of rolling body radius are calculated. Then we can enter the remaining data required to calculate forces and contact stresses (figure 4).

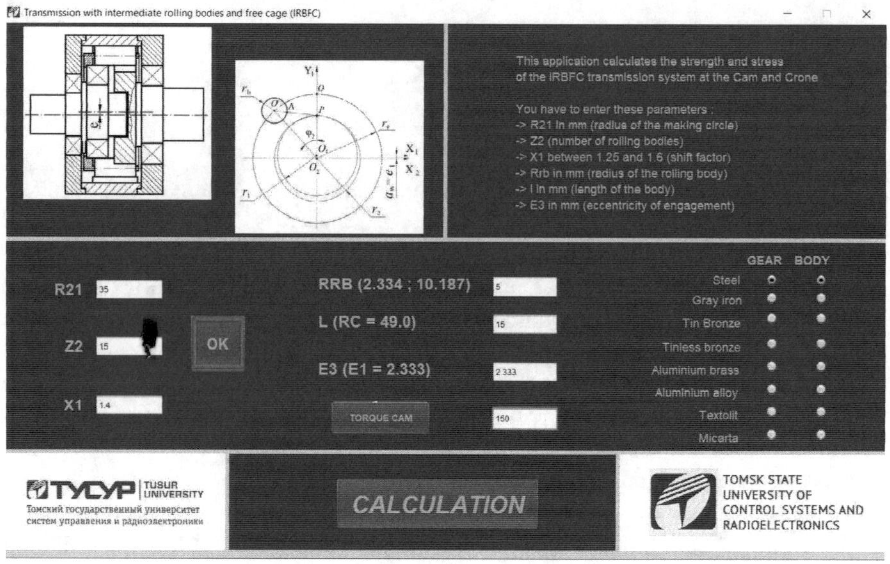

Figure 4. Introduction of initial and additional parameters for calculating forces and contact stresses.

When all the necessary data is entered and the material of the contacting transmission links is chose, the "CALCULATION" button is pressed, after which two windows are opened with graphs of the changes in the forces and contact stresses in the engagement of the transmission with IRBFC along the working length of the profile (figure 5).

One of the windows corresponds to the data by the cam wheel and the other window to the crown wheel. Moreover at the time of opening of these windows an excel file is created in parallel. This file contains all the values calculated and displayed by the windows. This file allows working directly with the exact values calculated by the program. The forces are expressed in Newtons and the contact stresses in Pascals. Since the accuracy of these data depends on the pitch of the angle φ (which varies from 0° to 180°), the chosen pitch is 0.5° which allows obtaining a good precision. Finally, we can see the data displayed in a drop-down table, as well as the maximum value of force and contact stress.

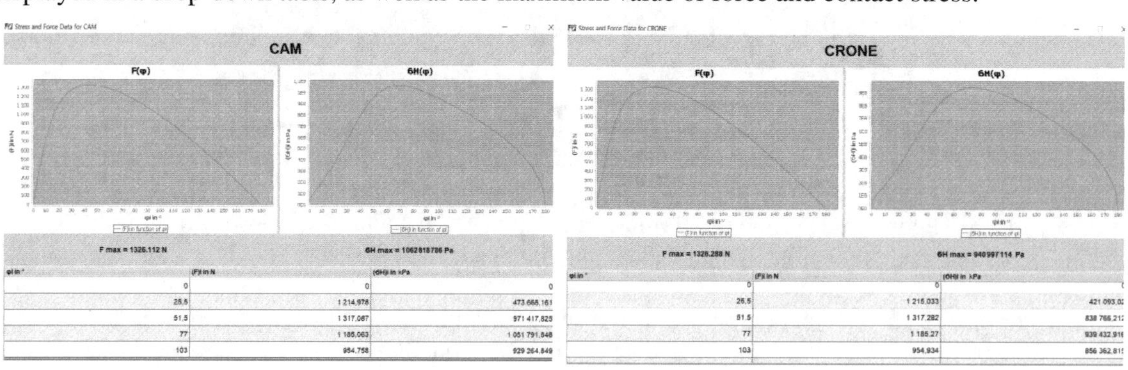

Figure 5. The results of the automated calculation program.

The creation of this computer program is unrolled in 2 parts:

-> the first part consisted of creating a simple program that performed all the calculations related to the mechanical system, and creating the mechanical transmission as a computer object (corresponds to a "class").

-> the second program was intended to create the graphical user interface (GUI), i.e. the link between the first part and the user of the computer program. Moreover, this part corresponds to the final program.

4. Example

Let us calculate of the transmission with rolling bodies from steel and gears from tinless bronze. We will make calculation for the transmission with the following initial parameters: R_{21}=35 mm; Z_2=15; χ=1.3.

After preliminary calculation we obtain: range of data for radius of rolling body – from 2.334 mm to 9.459 mm; eccentricity of cam and rolling body engagement – 2.333 mm.

We chose radius of rolling body R_{RB}=5 mm, and set torque on the cam like 200 Nm. Then calculation is continued. We obtain graphs of force and contact strength determination (figure 6).

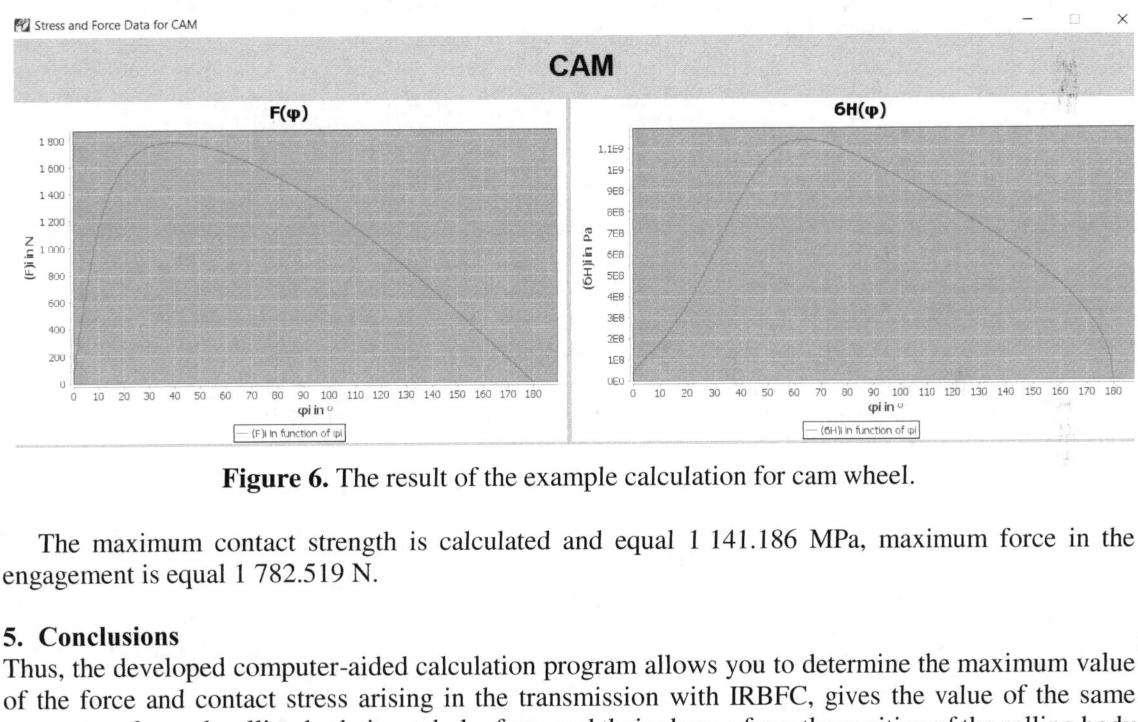

Figure 6. The result of the example calculation for cam wheel.

The maximum contact strength is calculated and equal 1 141.186 MPa, maximum force in the engagement is equal 1 782.519 N.

5. Conclusions

Thus, the developed computer-aided calculation program allows you to determine the maximum value of the force and contact stress arising in the transmission with IRBFC, gives the value of the same parameters for each rolling body in a tabular form and their change from the position of the rolling body in a graphical form. For the developed software, Certificate No. 2019660019 was received.

References:

[1] Pankratov E N 1998 Designing of mechanical systems of the automated complexes for mechano-machining of manufacture. A practical work of the leader-designer *Tomsk State University Press* p 296

[2] Aksenov V V, Efremenkov A B, Blaschuk M Yu and Ryltseva Ya G 2012 *J. Vestnik Nauki Sibiri (Messenger of Siberia Science)* **1 (2)** 372-378

[3] Lustenkov M E 2010 *J. Russian Eng. Res.* **30 (9)** 862-866

IOP Conf. Series: Materials Science and Engineering **795** (2020) 012025 doi:10.1088/1757-899X/795/1/012025

[4] Prudnikov A P 2018 *J. AER-Adv. in Eng. Res.* **158** 338–342

[5] Ephremenkov E 2006 Calculation of Temperature of Heating of Speed Reducers On The Basis of Transmissions with IRB *The 1st International Forum On Strategic Technology, Ulsan, Korea* 342–343

[6] An I-Kan, Il'in A S and Lazurkevich A V 2016 Load analysis of the planetary gear train with intermediate rollers. Part 2 *IOP Conf. Series: Materials Science and Engineering* **124**

[7] Lehmann M 1976 Calculation and measurement of forces acting on cycloidal speed reducer *PhD Thesis* Technical University Munich Germany

[8] Malhotra S K and Parameswaran M A 1983 Analysis of a cycloidal speed reducer *Mechanism and Machine Theory* **18 (6)** 491-499

[9] Yan H S and Lai T S 2002 Geometry design an elementary planetary gear train with cylindrical tooth profiles Mechanism and Machine Theory **37 (8)** 757–767

[10] Li X, He W, Li L and Schmidt L C 2004 A new cycloid drive with high-load capacity and high efficiency *ASME Journal of Mechanical Design* **126** 683–686

[11] Shin J H and Kwon S M 2006 On the lobe profile design in a cycloid reducer using instant velocity center *Mechanism and Machine Theory* **41** 596–616

[12] Blanche J G and Yang D C H 1989 Cycloid drives with machining tolerances, ASME Journal of Mechanisms *Transmissions and Automation in Design* **111** 337–344

[13] Blagojević M, Nikolić V, Marjanović N and Veljović LJ 2009 Analysis of cycloid drive dynamic behavior *Scientific Technical Review* **LIX (1)** 52-56

[14] Blagojević M, Marjanović N, Đorđević Z and Stojanović B 2011 Stress and Strain State of Single-Stage Cycloidal Speed Reducer *The 7th International Conference Research And Development Of Mechanical Elements And Systems IRMES-2011* 553-558

[15] Litvin F L and Feng P 1996 Computerized design and generation of cycloidal gearings *Mechanism and Machine Theory* **31 (7)** 891–911

[16] Efremenkov E A 2009 Determination of forces in transmission with intermediate rolling bodies and free cage *6th Intersectoral Scientific and Technical Conf. "Automation and Advanced Technologies in the Nuclear Industry", Novouralsk NSTI* pp 123-126

[17] Efremenkov E A 2002 Development of methods and means of increase of effectiveness of transmissions with intermediate rolling bodies: thesis on competition degree of candidate of technical science *Tomsk, TPU*

[18] E.A. Efremenkov and An I-Kan 2010 *Russian Engineering Research.* **30 (10)** 1001-04

IOP Conf. Series: Materials Science and Engineering **795** (2020) 012026 doi:10.1088/1757-899X/795/1/012026

Development of Energy Efficient Mechatronic Module for Alternative Energy

K G Shibinskiy, E A Efremenkov and N E Galin
School of Advanced Manufacturing Technologies, Tomsk Polytechnic University, 30 Lenin ave., Tomsk, 634050, Russia

E-mail: ephrea@mail.ru

Abstract. The article describes the creation of mechatronic module for follow-up systems. In the mechanical part of the module the authors applied a reducer, which is based on cam coaxial radial flat transmission with intermediate rolling bodies. Transmission ratio is determined and equation of tooth profile is derived for the transmission. Transmission kinematics with different number of rolling bodies was also analyzed. Mechatronic module was designed to limit overall dimensions. The analysis of stress-strain state of the most loaded elements of the transmission is conducted. Structural diagram of mechatronic module is developed. Structure of optimal control of BLDC motor is presented as well as a function of minimizing the dependence on characteristics of a transition process of motor speed is selected. An adaptive PID controller is synthesized using genetic algorithm.

1. Introduction

For mechatronic systems, the requirements of technical characteristics of modern mechanisms are applied. Gearmotors are one of the main driving devices of mechatronic systems. They are subject of high requirements in terms of mass-size-power ratio. The reducing mechanism is mainly responsible for providing this ratio. This ratio is best provided by rotation mechanisms made on the basis of kinematic wave transmissions [1, 2]. These transmissions receive a load of about half the teeth of the cycloid wheel, other words the teeth of the gear at an angle of approximately 180° are in the working area [3, 4]. This makes it possible to transmit large rotating moments relative to the toothed gears, with the same mass-dimensional characteristics. Modern drives use a compact motor with a hollow shaft like 3DBM (Figure 1). And in order to develop a mechatronic module based on this motor and cycloidal transmission with intermediate rolling bodies (IRB), it was a task to fit the reducer into the cavity of the motor shaft. This will minimize the overall dimensions of the mechatronic module and the entire mechanism; as well as develop an electric drive control system that allows to protect the drive from overheating and controlling the rotation speed of the motor shaft.

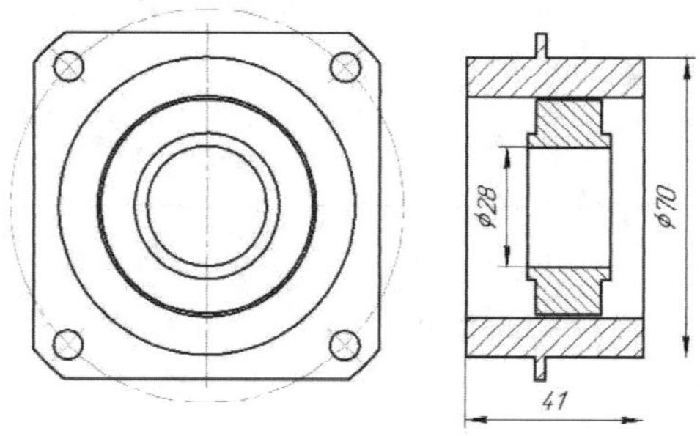

Figure 1. General view of the electric motor 3 DBM

Mechatronic module provides the following technical data:

- Reducer is designed for electric motor 3 DBM 70;
- Transmission ratio is 8;
- Rated torque, not less than 1 Nm;
- Maximum torque, not less than 1.5 Nm;
- Efficiency, not less than 85%;
- Weight of the reducer, not more than 0.25 kg.

There is a three-phase brushless direct current (BLDC) motor with the supply voltage of 27 volts, positioning accuracy on a turning angle 10 minutes, rotating speed of 3000 revolutions per minute that is considered as main characteristics of an electronic part of the module.

Currently, there are four generations of mechatronic modules. The most modern one is synergy of a mechanical part (the motor and reducer), an electronic part (drivers and microcontrollers) and control algorithms. The feature of fourth generation is application of optimal or adaptive control algorithms using methods of artificial intelligence.

2. Design of Compact Cycloidal Reducer

In order to minimize the dimensions of the reducer, we use a cam coaxial radial flat transmission with IRB (Figure 2), which is an analogue of the kinematic wave transmission.

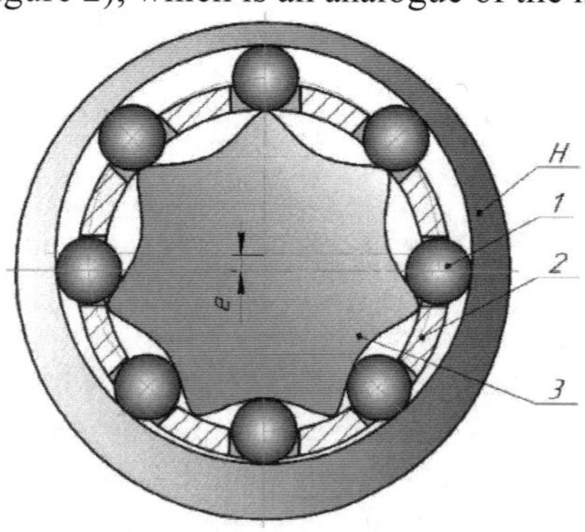

Figure 2. The cross section of cam coaxial radial flat transmission with intermediate roller bodies: *H* – generator (planet carrier); 1 – intermediate rolling body; 2 – separator; 3 – central gear;
e – eccentricity of the transmission.

The transmission works following way: the planet carrier *H* rotate and made intermediate rolling bodies to move radial along grooves of separator 2. The separator 2 is fixed rigidly in housing of the transmission. The rollers contact with cycloid profile of central gear 3. By the radial movement rollers made to the central gear 3 rotated.

There is no detailed description in the literature on the design of such gears, so it is necessary to determine the transmission ratio and infer the tooth profile equation for the gear.

In the transmission, rolling bodies have planetary rotation, i.e. this one can be attributed to epicyclic gear train [6]. Indeed, the wave generator can be called a planet carrier because it support the rolling bodies having movable axes: in the slots of the separator bodies reciprocate and rotate around their own axis. Thus, the Willis method is applicable to the present transmission [7]. This method based on the motion conversion (method of planet carrier stopping in the mind): all links of the mechanism are mentally given opposite rotation with angular speed equal to the angular speed of the planet carrier. Let us take wave generator with eccentric as planet carrier, rolling bodies with separator – link 2, profile gear – link 3. Since the separator is fixed, its angular velocity in the non-rotated mechanism will be equal to zero. We will draw up a table of links angular velocity for reverse and non-reverse transmission.

Table 1. Angular velocity of the transmission links

Mechanism	H	2	3
non-reverse	ω_H	0	ω_3
reverse	0	$-\omega_H$	$\omega_3-\omega_H$

Let us determine the transmission ratio from the third link to the second link in the reverse mechanism:

$$i_{32}^{(H)} = \frac{\omega_3-\omega_H}{-\omega_H}, \qquad (1)$$

By simplifying the expression (1), we get:

$$i_{32}^{(H)} = \frac{Z_2}{Z_3}. \qquad (2)$$

Transmission ratio from carrier H to link 3:

$$i_{H3} = \frac{1}{1-i_{32}^{(H)}}. \qquad (3)$$

Substituting (2) in (3) and converting, we get:

$$i_{H3} = \frac{Z_3}{Z_3-Z_2}. \qquad (4)$$

Equation (4) defines transmission ratio of cam coaxial radial flat transmission with free rolling bodies. It can be seen from the equation that the smaller the difference between rolling bodies number and the teeth number of the gear, the higher the gear ratio. The maximum transmission ratio can be obtained with the difference Z_2 and Z_3 per unity.

Maximum transmission ratio can be obtained in two cases:

$$Z_3 > Z_2 \text{ or } Z_3 < Z_2, \qquad (5)$$

Then (4) will take the form:

$$i_{H3} = Z_3 \text{ or } i_{H3} = -Z_3. \qquad (6)$$

It can be seen from the expressions (6) that the rotation can be organized unidirectional or different-directional for input and output shafts in the transmission.

Let us get the equation of the gear tooth profile. For this purpose we will replace of the mechanism with higher kinematic pairs by the mechanism with lower pairs [8]. Then the task of the transmission kinematic investigation can be reduced to the corresponding equivalent axial slider-crank mechanism investigation [7] (Fig.3).

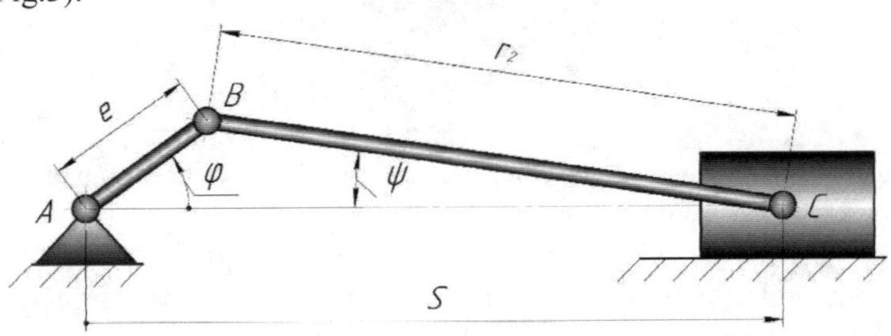

Figure 3. Replacement axial slider-crank mechanism

The input eccentric generator acts as the driving crank (Fig.3). Driving crank length is equal to generator eccentricity e. Connecting rod presents like line is connecting centers of rolling body and eccentric. Let us name the length of the connecting rod as the radius of the rolling body centers.
Distance S (Figure 3) can be written in vector form:

$$\vec{S} = \vec{e} + \vec{r_2} \tag{7}$$

Let us define the angle using the sine theorem

$$\frac{r_2}{\sin\varphi} = \frac{e}{\sin\psi}. \tag{8}$$

or

$$\psi = \arcsin\left(\frac{e}{r_2}\sin\varphi\right). \tag{9}$$

The displacement $\vec{S}$ formula in scalar form is written as:

$$S = r_2\cos\psi + e\cos\varphi. \tag{10}$$

where e – eccentricity of the transmission, value of which is recommended to be determined depending on radius of rolling body [9].

The expression (10) determines the position of the center of the rolling body relative to the transmission axis in the fixed coordinate system. It is the main geometric expression from which all another geometric equations are derived [9]. The equations of the centers of the rolling bodies, relative to the center of the gear, using the transition matrix [10] will take the form:

$$\begin{cases} X_2 = S \cdot \sin\beta; \\ Y_2 = S \cdot \cos\beta. \end{cases} \tag{11}$$

where β – rotation angle of circle of centers of rolling bodies relative to generator.

This angle is related to the generator rotation angle by the following constraint:

$$\beta = \frac{\varphi}{i_{H3}}. \tag{12}$$

Let us define the theoretical profile of the gear that point K of the rolling body surface will draw (Figure 4).

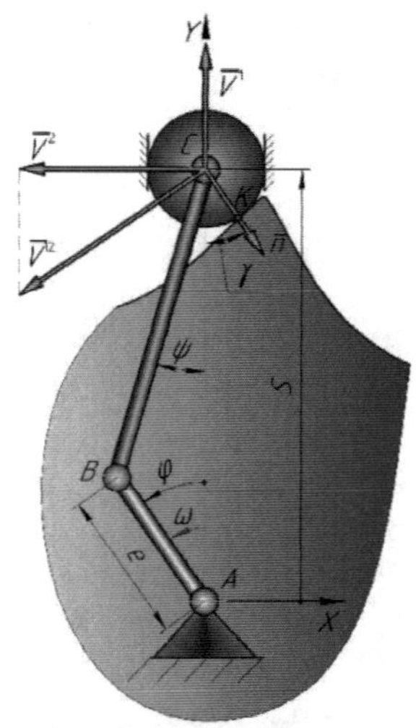

Figure 4. To define normal at contact point of rolling body and tooth profile

The vector $\vec{n}$ is normal of tangent to the gear profile at the point of contact with the rolling body. In addition, it is just as normal to the relative velocity vector $\vec{V}^{12}$. In order to determine the normal vector $\vec{n}$, we will decompose the velocity vector into components: radial $\vec{V}^1$, and tangential $\vec{V}^2$. The vector $\vec{V}^1$ of radial velocity can be determined by derivative (10) over time. The radial component of the relative velocity can then be determined as:

$$V^1 = \frac{\partial S}{\partial t} = -r_2 \sin\psi \frac{\partial\psi}{\partial t} - e\sin\varphi \frac{\partial\varphi}{\partial t}. \tag{13}$$

Differentiating by time (8) and setting the result to (13), we get:

$$V^1 = \frac{\left(-r_2 \sin\psi \frac{e}{r_2} \cos\varphi\right)\omega}{\cos\psi} - e\sin\varphi \cdot \omega. \quad (14)$$

The expression (14) is simplified, let us present it as an analogue (i.e., change by angular velocity), we obtain:

$$V^1 = -e\,\mathrm{tg}\,\psi \cos\varphi - e\sin\varphi. \quad (15)$$

Tangential component of relative speed $\vec{V}^2$ is determined through distance S (Figure 4) and angular velocity of input link. So as an analogue:

$$V^2 = \frac{S}{i_{H3}}. \quad (16)$$

Let us determine pressure angle γ [8] through relative velocity components:

$$\gamma = \mathrm{arctg}\frac{V^1}{V^2}. \quad (17)$$

The coordinates of the point *K* are defined through the radius-vector *R*, which can be found from the expression:

$$R = S\boldsymbol{j} + r = S\boldsymbol{j} + (r_{tk}\sin\gamma)\boldsymbol{i} - (r_{tk}\cos\gamma)\boldsymbol{j}, \quad (18)$$

where j – unit vector of axis Y;
i – unit vector of axis X.

By simplifying the expression (18) and bringing it into matrix form, we get:

$$R = \begin{pmatrix} r_{tk}\sin\gamma \\ S - r_{tk}\cos\gamma \\ 1 \end{pmatrix}. \quad (19)$$

The expression (19) is the coordinates of the point *K* in the fixed coordinate system. With the help of transfer matrix [10] we pass to coordinate system rigidly connected to rotating profile gear. Then the gear profile equation will take the form:

$$\begin{cases} X_3 = r_{tk}\sin(\gamma - \beta) + S\sin\beta \\ Y_3 = -r_{tk}\cos(\gamma - \beta) + S\cos\beta \end{cases}. \quad (20)$$

Equations system (20) defines parametric equation of gear profile in fixed coordinate system.

Equations of gear profiles at rotation opposite direction of output link (6) taking into account that sine of angle is even function, and cosine is odd function [11], will take the form:

$$\begin{cases} X_3 = -(r_{tk}\sin(\gamma - \beta) + S\sin\beta) \\ Y_3 = -r_{tk}\cos(\gamma - \beta) + S\cos\beta \end{cases}. \quad (21)$$

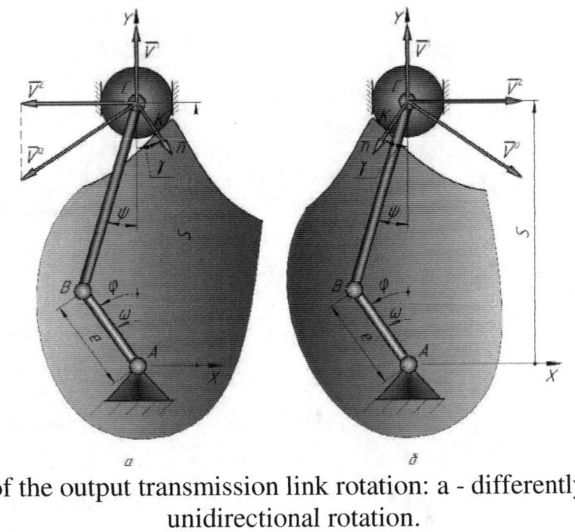

a б

Figure 5. Direction of the output transmission link rotation: a - differently directed rotation; b - unidirectional rotation.

It can be seen from the equations (20) and (21), when the direction rotation of the output link is changed, the coordinate *Y* of the gear profile remains unchanged and the coordinate *X* changes its sign to the opposite. At the same time rolling bodies will transmit load either right part of the profile or left part (Figure 5). Using the equations (20) and (21) it is possible to manufacture profile wheels by various methods [12, 13].

3. Development of the mechatronic module electronic part

As in the set BLDC motor there are no built-in current sensors, for control of speed of motor shaft the AS5147P sensor is used. It has 4096 values on one turn of a shaft that gives definition accuracy less than 6 minutes. This sensor is powered from 3.3 V and issues data on a shaft turning angle on the serial peripheral interface (SPI). For definition of a turning angle of a reducer shaft, the sensor of absolute situation is installed. It transfers data by industry standard RS422 to signal of universal asynchronous receiver-transmitter (UART) is transformed for processing by the microcontroller. For motor management the driver of STM L6235Q was selected, management of which requires 7 exits (3 PWM and 4 digital) and maintaining voltage to 52 V and current of 5.6 A. The industry standard is used RS485 for the set of the necessary rotational speed of the motor and a turning angle of the reducer shaft. The standard is transformed to the UART interface for interaction with the controller. The controller of STM STM32F407GT was selected as a result of the choice of these devices, as the managing microcontroller. The described system provides:

- data processing from the user and from the sensor of absolute position (two hardware UART interfaces);
- motor management (through 4 digital and 3 hardware PWM exit);
- data processing from the motor position sensor (the hardware SPI);
- operation frequency of the microcontroller – 64MHz;
- compact sizes.

The block diagram of the mechatronic module (figure 6) is a result of development.

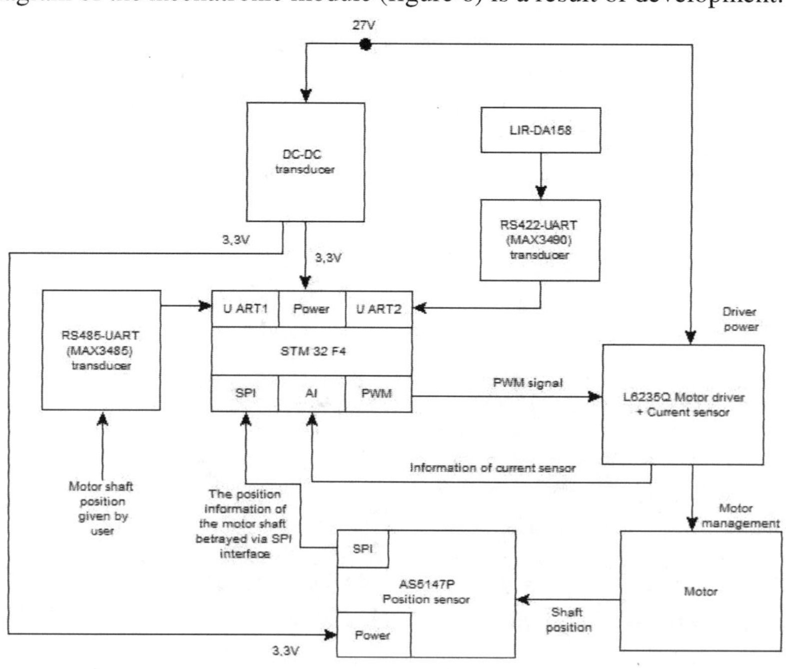

Figure 6. Structural diagram of the drive control system

To adjust the drive, a mathematical model BLDC motor was taken, it presented in [14]. During initial adjustment of proportional integral derivative (PID) controller the following factors were obtained: P = 0.33678, I = 37.5306, D = 0.0007489.

Since it is very difficult to establish a clear functional relationship between the indicators of the quality of the transition process and the coefficients of the PID controller, the authors preferred the use of artificial intelligence methods, which showed their validity both in the configuration of the PID controller parameters [15] and in the management of complex objects [16].

4. Example

On the basis of the data from the technical assignment, transmission parameters with intermediate rolling bodies were selected, the dimensions of which are located in the hollow part of the rotor (Fig.1). Transmission parameters are given in Table 2, the diagram of reduction gear box assembly is shown in Figure 7.

Table 2. Angular velocity of the transmission links

Parameter name	Designation	Value
Number of rolling bodies	Z_2	8
Rows number of rolling bodies	–	4
Radius of the rolling body centres, mm	r_2	7.925
Rolling body radius, mm	r_{tk}	1,6
Eccentricity, mm	e	0,8
Internal radius of generator, mm	–	9.525
External radius of generator, mm	–	14
One row width of rolling bodies, mm	–	4,3

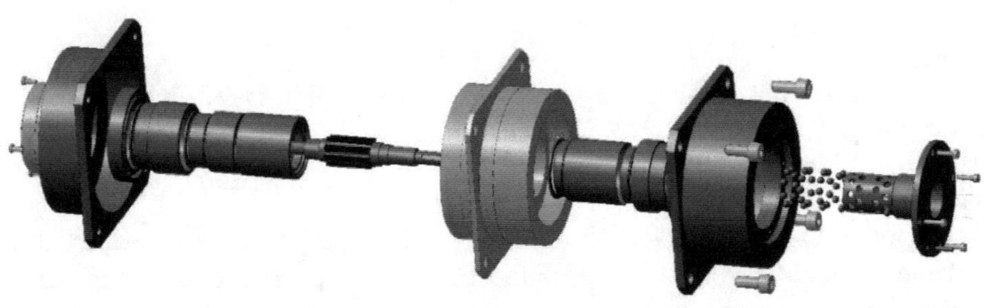

Figure 7. Assembly diagram of mechatronic module

Results of stress-strain state mathematical simulation are shown in Figure 8 for separator and the gear. For the most loaded parts of the reducer simulation was carried out in the SolidWorks system with maximum torque at 1.5 Nm.

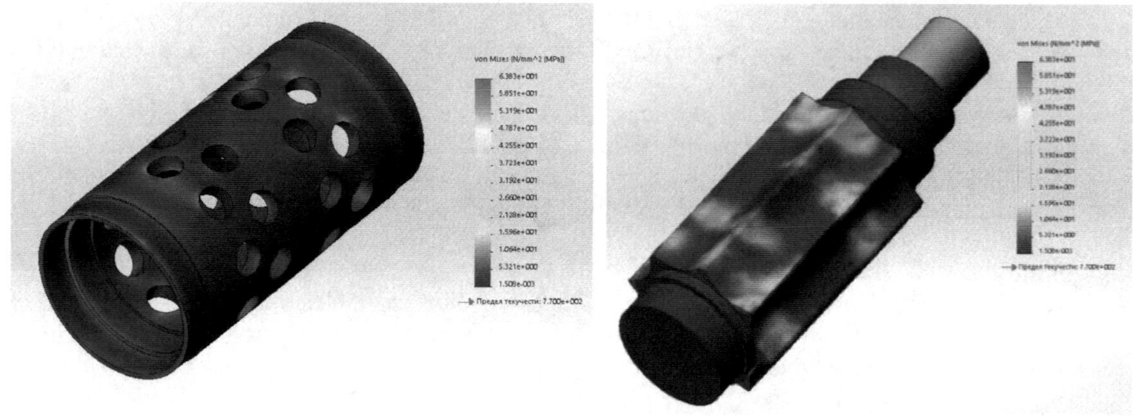

a) b)

Figure 8. Stress distribution in parts of reducer: a - separator; b - gear profile

5. Conclusions

On the basis of the electric motor of 3 DBM 70-0.16-3-3 with dimensions of 70x81x81 mm and weighing about 900 grams, the mechatronic module is developed.

It was possible to reduce overall dimensions of mechatronic module due to installation of coaxial reducer with intermediate rolling bodies into cavity of electric motor rotor.

Transmission ratio of cam coaxial radial flat transmission with IRB is determined and equation of profile is derived for the gear.

Intelligent control system of electric drive is proposed which to protect the module from overheating and optimize rotation speed of motor shaft by means of using PID controller with self-tuning function based on genetic algorithms.

Acknowledgments

The work was carried out with the financial support of the Ministry of Education and Science of the Russian Federation (Contract No. 02.G25.31.0190).

References:

[1] Stepanova D L, Nikulin P A and Krauin'sh P Ya 2016 Research of load distribution of wave kinematic reducer with modified tooth profile and dependence of loading capacities relative to its velocity ratio and overall dimensions *Internet-journal Naukovedenie* 8 4(35) [in Russian]

[2] Yali Yi, Yunfei Gao, Rui Wei and Herong Jin 2019 Teeth profile parameter design and transmission property analysis of external generating wave movable teeth transmission *The J. of Engineering* **13** 299-305

[3] Lustenkov M E 2010 *J. Russian Eng. Res.* **30** 9 862-866

[4] Efremenkov E A 2006 Calculation of temperature of heating of speed reducers on the basis of transmissions with IRB *The 1st International Forum On Strategic Technology "e-Vehicle Technology", IFOST-2006* pp 342-343

[5] Shatalov E V, Efremenkov E A and Shibinskiy K G 2016 Determination of Radius of Curvature for Teeth With Cycloid Profile *IOP Conf. Series: Materials Science and Engineering* **124** 1 012007.

[6] Kudryavcev V N 1966 *Epicyclic gear trains* (Moscow: Mashinostroenie) **308**

[7] Efanov F M and Kovalevskiy V P 2004 *Mechanism and machine theory* (Orenburg: OSU Press)
[8] Fedotov G D 2013 *Mechanism and machine theory* (Ulyanovsk: USAA named after P.A. Stolypin Press)
[9] Belyaev A E 1986 *Mechanical transmissions with intermediate bodies of increased accuracy and durability* (Tomsk: TPI Press)
[10] Litvin A F 1960 Tooth system theory (Moskow-Leningrad: Fizmathgiz Press) p **585**
[11] Bronshtejn I N and Semendyaev K A 2010 Mathematics handbook for engineers and students in the back: manual / /mathematics 51 08 [in Russian]
[12] Ambrosimov S K and Khandozhko A V 2017 Metal cutting of tooth gears by a method of a nonlinear generating process of a profile *IOP Conf. Series: Materials Science and Engineering* **177** 1 012050
[13] Ambrosimov S K and Morozova A V 2018 Investigation of formation of cut off layers and productivity of screw milling process *IOP Conf. Series: Materials Science and Engineering* **327** 4 042003
[14] Yu G and Hwang R 2004 Optimal PID Speed Control of Brushless DC Motors Using LQR Approach *IEEE International Conf. on Systems, Man and Cybernetics* pp 473–478
[15] Zhang S, Zhou X and Yang L 2011 Adaptive PID regulator based on neural network for DC motor speed control *International Conf. on Electrical and Control Engineering* pp 1950–1953
[16] Mamonova T, Syryamkin V and Vasilyeva T 2016 Modelling of the automatic stabilization system of the aircraft course by a fuzzy logic method *IOP Conf. Series: Materials Science and Engineering* **124** (1)

IOP Conf. Series: Materials Science and Engineering **795** (2020) 012027 doi:10.1088/1757-899X/795/1/012027

Factors influencing strength and direction of natural ventilation pressure in general mines to control mine airing

G B Lyalkina, A V Nikolaev, N S Makarychev

Perm National Research Polytechnic University, 29 Komsomolskii av., Perm, 614990, Russian Federation

E-mail: nikolaev0811@mail.ru

Abstract. The article presents three main factors, influencing the strength of natural ventilating pressure in general mines between the shafts: gas emission from mines (air density changes due to the change in concentration of other gases), changes of air parameters taken in by ventilating shafts, and air pressure due to the productivity shifts of the Main Fan Unit (or other devices for positive and negative control of air distribution). According to the experiment data the equations mentioned in other works for calculating natural ventilating pressure in general mines and taking into account the parameters of the air taken in are precise. Applying these formulas basing on the results of hydro meteorological forecast, whose accuracy in the short-term forecast is also high, it is possible predict the operating mode of the system of ventilation.

1. Introduction

During the underground mining it is necessary to ventilate the mines to provide safe working conditions for people. Main Fan Units (MFU) are applied in mines and shafts for this. They feed air down to shafts and mines into working zones. The number of production units, number of workers in a mine (shaft), gas volume and other factors determine an amount of air necessary to ventilate the underground mine. These factors determine the required capacity of a MFU. Ventilating practice shows that the volume flow of the air fed into the mine (shaft) changes constantly. One of the factors influencing significantly the ventilating process in underground mines is natural ventilating pressure in general mines between the shafts. When the natural ventilating pressure coincides with required direction of the air flow in the shaft it is called a positive natural ventilating pressure in a general mine. If it is reversed to the required air flow it is called a negative natural ventilating pressure and hinders the required ventilating process [5, 6].

Many works have been devoted to mathematical functions for determining the volume and direction of the natural ventilating pressure. M. V. Lomonosov was the first to pay attention to the natural ventilating pressure in mines. The first mathematical dependencies for determining the absolute values of the natural ventilating pressure in mines were introduced in [7-9]. The authors took into account the state of the atmosphere in a mine (shaft) which was considered in calculations as polytrophic, homogenous, isothermic and adiabatic, as well as mine depth and pressure developed by the MFU. Works [10, 11] take into account the influence of the natural ventilating pressure on air distribution in the system of ventilation. Authors of the consequent works analyzing the influence of different factors on the value of the natural ventilating pressure suggested both general and simplified formulas. For example, in [12] the authors considered that the natural ventilating pressure does not depend on the mine depth and MFU operation. In the article [13] it is stated that in the mine with a single level shaft the natural ventilating pressure is initiated only by forced air supply that is with the help of so called "air acceleration". These articles also state that the natural ventilating pressure increases with the shaft depth due to the air temperature increase and increases with the mine (shaft) depth.

Thus, we can state that nowadays, there is no common point of view on the appearance of natural ventilating pressure and factors influencing its magnitude and direction.

2. Results

Nowadays, there is no scientific explanation of the reasons of natural ventilating pressure occurence and regularities of changes in it, as well as its influence on the operation of MFU, which makes impossible to create the methodology of monitoring the ventilating process of underground mines with the account of its operation.

It can be noted that two factors influence the natural ventilating pressure – air density and the variation between the spot height of the interconnected mines. We can make a conclusion that the value of the natural ventilating pressure between the interconnected mines is influenced by the spot heights which do not change relative to each other (for example, the shafts), only the air density will have influence in these mines.

Three groups of reasons influence the changes in air density, consequently, the magnitude of natural forces causing the displacement of air flows (masses):

1. Changes in air density at constant temperature and pressure, due to changes in gas concentration in the mine.
2. Air density changes due to its temperature fluctuation. The appearing draft in this case is called a "thermal drop of ventilation pressure".
3. Changes in air density due to areas with increased or decreased pressure, which appear due to many reasons, such as: collision of air flows, redistribution in MFU reverso or changes in its capacity, when air flows runs onto artificial obstacles (shaft air bratices, air doors and so on).

Thus, to determine the factors influencing the value of the natural ventilating pressure it is necessary to study separately the influence of these parameters on its value.

The article [14] provides the results of studying the changes in natural ventilating pressure values depending on the gas concentration appearing while mining and influencing the air density in the underground space. The authors conclude that the mechanism of natural ventilating pressure occurence between the underground mines will be different for gas-bearing and non-bearing mines. The authors call this effect "gas ventilation pressure". Mathematical functions have been recieved of density difference dependence of "pure" air and gas saturated air on the value of natural ventilating pressure. The studies have been carried out for the air standstill in the mine, when the pressure changes (consequently, air density) is due to changes in concentration, that is when the air temperature is constant. It can be stated that this article proves the fact of changes in natural ventilating pressure when gases of the different density appear from the geological material (for example, methane). For obvious reasons this factor is unique to gas bearing mines and shafts.

To determine the average air density in shafts versus its temperature and pressure, the paper [5] introduces formulas. However, it was imposible to determine their accuracy till nowadays. It is caused by the following challenge.

The ventilating process is slow responding – after shifts in MFU productivity air capacity in downcast ventilating shafts will change only in some time [15]. After measuring air parameters in airfeeding and ventilating shafts it is impossible to determine the reasons of air distribution changes due to the difference of air capacity in the shafts. Due to this the article [16] suggests the calculation method of the value and direction of the natural ventilating pressure measured at MFU. Here, when MFU productivity changes the values of natural ventilating pressure (h_e) and air pressure of the mine (R_{mine}) are determined by the least squares method the performance curves are built with and without the account of this pressure.

The article gives the results of calculating natural ventilating pressure in general mines in summer and in winter with this method, the acuracy of the recieved data and their comparison has been done with the calculated ones according to operational formulas [5]. The experiments have been conducted at the mine PJSC "Uralkali").

The calculation data by the method [16] of the air pressure of the mine (R_{mine}) with the account of (lines # 1 and 2) and without the account of (lines № 3 and 4) natural ventilating pressure (h_e) are in table 1.

Table 1. The results of calculating the air pressure of the mine and natural ventilating pressure in general mines according to the experiment results (when MFU productivity changes)

	The season	Parameter			
		R_{mine}, $(H{\cdot}s^2)/m^8$	h_e, Pa	h_{lower}, Pa	h_{upper}, Pa
1	Summer experiment	0.01374	406.1	386.77	425.44
2	Winter experiment	0.01686	75.99	66.80	85.18
3	Summer experiment	0.01745	–	–	–
4	Winter experiment	0.01620	–	–	–

Fig. 1 shows the mine curves by the determined values R_{mine} and h_e: with the account of natural ventilating pressure (curve 1) and without it (curve 2). Figure 1 shows experimental points of the measured MFU operational modes at different speeds of rotor rotation (curve 3).

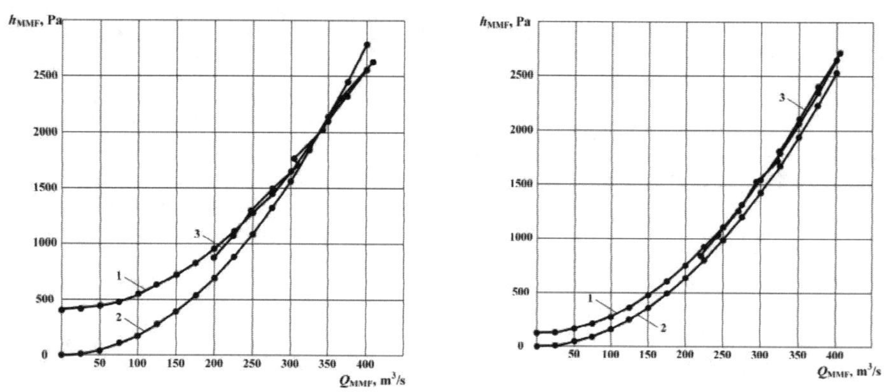

a) *b*)

Figure 1. The graphs constructed according to the data with the account of natural ventilating pressure and without it, as well as the results of the experiment: *a* – in summer; *b* – in winter

As graphs on Fig. 1 show the experimental points have more coincidents with the curves 1, that is with the curves, built with the account of the natural ventilating pressure in general mines. Here we can see that the experimental data, recieved in winter are not identical to the mine characteristic 2 as data in summer. In this case with the fudicial probability we can suggest that in both experiments the air pressure of the mine h_e appeared but in winter its value was little. Table 1 shows that the values of the natural ventilating pressure were as follows: in summer $h_e = 406.1$ Pa, in winter h_e: = 75.99 Pa.

3. Discussion

Taking into account the fact that the method of determininng natural ventilating pressure at changes on MFU it is necessary to compare the recieved value of h_e with the recieved by formulas of operation [5] (let's call them theoretical formulas). For this the parameters of the air outside were introduced into the formula: temperature and air pressure. As in winter the air was dried (mine air heating installations), the air temperature in downcast ventilating shafts was measured by sensors inside them.

The following data have been recieved.

In summer the value of natural ventilating pressure was $h_e = 387.3$ Pa, in winter – $h_e = 64.9$ Pa.

Comparing the values he, recieved in winter and in summer with the reieved by theoretical formulas values we can see that their variation is: 4.85% (summer) and 16.95% (winter).

In summer there the experimental points almost coincided with the characteristic curve of the mine built with the account of natural ventilating pressure (curve 1, Fig. 1, *a*). At this moment we can say that the natural ventilating pressure existed and its value was almost the same as recieved. The difference between the experimental value of the natural ventilating pressure in general mines and the one recieved by theoretical formulas is less than 5%.

The difference 16.95% is not so great, but we can suggest that this difference is due to poor convergence of the experimental data with the curve 1 (Fig. 1, *b*). Here, the winter data h_e can have some error. Perhaps this error influenced the difference of natural ventilating pressure values in winter and according to theoretical formulas.

In any case we can say that the theoretical formulas of operation [5] with high precision can be used for calculating the value and direction of the natural ventilating pressure.

4. Conclusion

The study has shown that to determine the value and direction of natural ventilating pressure it is necessary to take into account a gas presence in a mine. Changes in air density due to changes in gas concentratons in it will influence the natural ventilating pressure between the interconnected mines.

For mines with low volume of gas it has been determined that the value of natural ventilating pressure h_e with the high degree of accuracy can be determined by the derived formulas of operation [5], which can be separately sed to determine the values of pressure and air temperature. Here we can determine the influence of these air parameters on the value of natural ventilating pressure separately.

It can be noted that the air parameters from hydrometeorologic forecasts can be introduced into these formulas. It has been proved that the accuracy of these forecasts nowadays is 85-90% [17, 18]. Consequently the obtained data can be used to calculate the predicted value of natural vntilating pressure, that allows to choose the required mode of MFU operation and the system of air drying beforehand (if it is necessary). It will allow controlling the ventillating process in a safe and energy saving mode.

References

[1] Bruce W E 1986 Natural draft: its measurement and modeling in underground mine ventilation systems. US: Dept. of Labor. Mine Safety and Health Administration.

[2] Linden P F 1999 The fluid mechanics of natural ventilation *Annual Review of Fluid Mechanics* **31** 201–238

[3] Jianwei Cheng, Yan Wu, Haiming Xu, Jin Liu, Yekang Yang, Huangjun Deng and Yi Wang 2015 Comprehensive and integrated mine ventilation consultation model *Tunneling and Underground Space Technology* **45** 166–180

[4] Gendler S G 2016 The justification of new technique ventilation at contraction of working with two exits in soil surface *Eurasian Mining* **2** 41–44.

[5] Alymenko N I and Nikolaev A V 2011 Influence of mutual alignment of mine shafts on thermal drop of ventilation pressure between the shafts *Journal of Mining Science* **47 5** 636–642

[6] Nikolaev A V, Alymenko N I, Kamenskikh A A, Alymenko D N, Nikolaev V A and Petrov A I 2017 Factors defining value and direction of thermal drop between mine shafts and impact of general mine natural draft on ventilation process of underground mining companies *Actual Issues of Mechanical Engineering (AIME 2017)* 561–566

[7] Komarov V B and Borisov D F 1938 *Mine ventilation.* (L.: NCSTPH) p 454

[8] Borisov D F 1941 Determining the drop of natural ventilating pressure by the calculation method (St.-Petersburg notes) **14** 87–108 [in Russian]
[9] Voropaev A F 1949 Thermodynamic method of determining drop in natural ventilation pressure in mines *Ore and Metals* **7** 21–26
[10] Vasserman A D, Alehichev S P and Krotov K P 1966 Calculation of air distribution in a ventilating system under a complex natural ventilating pressure *Ventilating the shafts and mines with vast caved zones* 60–66 [in Russian]
[11] Alekhichev S P and Kalabin G V 1974 Natural ventilating pressure and the termal environment in mines (L.: Nauka) [in Russian]
[12] Kazakov B P, Shalimov A V and Semin M A 2015 Stability of natural ventilation mode after main fan stoppage *International Journal of Heat and Mass Transfer* **86** 288-293
[13] Karelin V N, Kravchenko A V, Yu L L, Kazakov B P and Zaitsev A V 2013 Features of forming microclimatic conditions in mining excavations of deep mines *Gornyi Zhurnal* **6** 65-68 [in Russian]
[14] Aitao Z and Wang K 2018 Role of Gas Ventilation Pressure on the Stability of Airway Airflow in Underground Ventilation *Journal of Mining Science* **54** **12** 111–119
[15] Lyal'kina G B, Nikolaev A V and Makarychev N S 2018 Creation of the Information System Based on Experimental Data for Control of the MMF Operating Modes to Improve the Efficiency of Ventilation in Mines *IOP Journal of Physics: Conf. Series* **1059** 012013
[16] Lyal'kina G B and Nikolaev A V 2015 Natural draught and its direction in a mine at the preset confidence coefficient *Journal of Mining Science* **51** **2** 342–346
[17] Tsirkunov V, Ulatov S and Korshunov A 2004 Assessment of economic efficiency of the National hydrometeorological system modernization project. World Bank Working Paper.
[18] Shaimardanov M Z and Korshunov A A 1994 The use of hydrometeorological information in the various economic sectors *Conference on the economic benefits of meteorological and hydrological services*. Geneva, WMO/TD **630** 28-36

IOP Conf. Series: Materials Science and Engineering **795** (2020) 012028 doi:10.1088/1757-899X/795/1/012028

Designing an Information Security System in Smart Factories of the Industry 4.0

D A Zakoldaev[1], A V Shukalov[1], I O Zharinov[1] and O O Zharinov[2]

[1]Faculty of Information Security and Computer Technologies, Saint Petersburg National Research University of Information Technologies, Mechanics and Optics, 49, Kronverksky Av., Saint Petersburg, 197101, Russia

[2]Department of Problem-Oriented Computing Complexes, Saint Petersburg State University of Aerospace Instrumentation, 67, Bolshaya Morskaia str., Saint Petersburg, 190000, Russia

E-mail: mpbva@mail.ru

Abstract. The problem of designing an information security system in smart factories of Industry 4.0 is investigated. Information security system is a company infrastructure component to provide the company information security. In the example of typical smart factory structure cyber and physical components are defined which are mostly susceptible to the malicious software components implementations. Such components could be components with their own means of calculations and engaged in the production technological operations. Smart factory information security risks can be explained with program and technical means for the operator to control the cyber and physical configuration remotely and other components of the production infrastructure into a single digital space of the company. Measures are proposed to increase the information security level of the Industry 4.0 smart factory based on technical, organization and program solutions.

1. Introduction

The wide application of advanced informative technologies and digital cyber and physical systems in the Industry 4.0 [1, 2] companies provided a background for attackers specialized in unauthorized access to the production components and data rights obtaining [3]. Their special objects are separate cyber and physical systems (CPS), technological lines, production and finance data to characterize the company activity and some other components of the Industry 4.0 cyber and physical production, which are of interest for wicked competitors [4].

According to the tasks, the attackers could see their targets as follows [5, 6]:

- acquisition of technical and economic data on project and production company activity;
- intentional failure creation of some company production infrastructure components including the error of data access or integrity (technical documentation);
- temporal or constant unauthorized access to the company production components and other.

There are several ways for an attacker to hack the company infrastructure production components [7, 8]. In digitalizing project and production activity in most cases, the attackers implement a malicious software into company informative system. To prevent an attacker unauthorized access to the production infrastructure they must develop and implement into the company security system special organization, technical, program and other measures which together provides the company information security [9 ,10].

2. Cyber and physical production as an unauthorized access subject

Cyber and physical production information security is based on some scientific and practical tasks solution, which may include [11]:

- production division structure analysis and sorting out the components, which are potential targets for an attacker;
- detection of threats and risks, which are important for the company if the attacker unauthorized access to the production infrastructure components will be realized using one of the methods;
- development and research of the attacker model to understand its target, objectives, methods and means, which are helpful for an unauthorized access to the company production infrastructure granting;
- development, test and implementing to the company information security system the measures, methods and means reducing the probability that an attacker would be a threat to the production division.

Figure 1 there is a scheme of cyber and physical production division as an attacker unauthorized access subject. The main attention is paid to the objects detection, which are potentially vulnerable for an implementation of a malicious software.

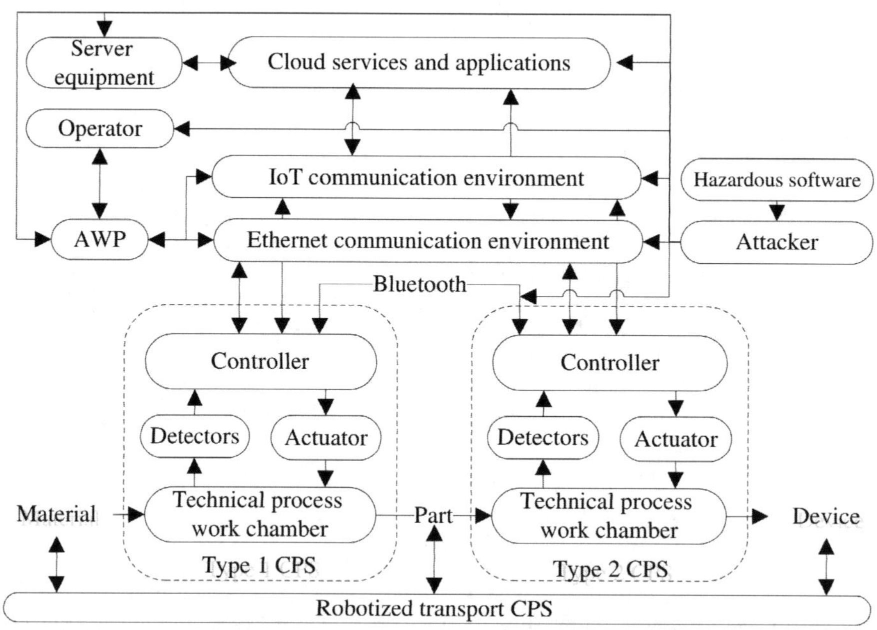

Figure 1. Cyber and physical production scheme as an unauthorized access subject (AWP – Automatic Work Place, IoT – Internet of Things).

The following groups of the company production infrastructure components could be susceptible to the unauthorized access threat [12]:

- cyber and physical production machines with their own calculation means are capable to register, store and execute the malicious software;
- server equipment with its own calculation resources is the company virtual environment based on which cloud services and applications are realized to provide CPS functionality;
- communication devices (phone, PC, tablet and other) grant to an operator the access to the production division components with its own operational system which is capable to execute potentially malicious software.

Cyber and physical production objects unification into a single technological line requires creating the information security management system. The information security control is the procedure to provide automatic and continuous company production resources functionality, which may include:

- technical and program means of cyber and physical systems;
- technical and program means of the company virtual environment;
- production data base and cyber and physical production base of knowledge to provide information accompanying technological operations processes completion and other.

The information security management system of a company is placed in virtual environment of the company and functions based on production data transmitting communication environment [13].

3. Communication environment as a way of malicious software distribution

Cyber and physical components interaction is done with the production data information exchange in the company communication environment. Alongside with the good function of the company communication environment, there could be a potential information channel capable to distribute (to implement into the production components) malicious software of an attacker.

Cyber and physical production communication environment is formed by wired and wireless communication channels. The Ethernet wired link channel unites CPSs and company virtual environment servers. The IoT wireless channels and Bluetooth unite CPSs and grant them access to the cloud services and applications. CPS as an active smart factory production infrastructure Internet agent provides remote information data exchange with a computerized control system (For example, CPS messages about components and materials orders which are necessary to complete technological operations) which is a potential risk of malicious software distribution through the channels of information transmission within the entire company.

The attacker actions model who is interested to get right of an unauthorized access to the production components may include the following ways of malicious software distribution:

- connection to the external communication devices IoT channel, which contains the malicious software to place it in the production components;
- malicious software distribution to the production operator communication devices to implement the virus secretly into the production infrastructure;
- malicious software distribution to the company virtual environment server equipment through the channels of remote access through the Internet;
- malicious software distribution to the CPS controller with the Bluetooth channel in the remote control mode (servicing) of cyber and physical system with the following software distribution in the IoT net to all production components;
- connection to the Ethernet net to distribute in any available production component a malicious software;
- designer implementation to the program code of cloud services and applications a piece of malicious software which will be further distributed to the net of Ethernet and IoT to the cyber and physical system;
- CPS designer implementation to the integrated software a piece of malicious software code which permits an attacker to control remotely (say, to switch off) a CPS and other.

Any of those attacker options to implement a malicious software to the production components leads to the quick distribution of the virus and causes the breakdown of the company control computerized system. Nowadays to detect those software components or attempts to implement them, the company should create machine and program means of protection, which are meant to be stationary installed in the production components but also they could be mobile which are used in the smart factory control technological processes.

The cyber and physical production communication environment organization scheme based on heterogeneous nets apart from threats of information security in the same net is susceptible to the threats arisen from distributed character of production resources and data. Local nets of Ethernet and Bluetooth of cyber and physical production are of attacker's interest to receive the rights of unauthorized access to the production data and to violate the production data integrity.

In the industrial Internet of Things, there is also the threat for company information security net such as blocking the message transmission between CPSs and cloud services and applications, as well as possible failure of CPS servicing (CPS data accessibility). The global IoT net threats are normally remote.

A cyber and physical production must have preventive measures against the following:

- an unauthorized connection to the company net;
- an unauthorized corrections of the cloud services and applications;
- an unauthorized protocol changes of production data transmission;
- an unauthorized changes of identification order and authentication participants of production process and other.

The company information security task is further complicated by the fact that all types of technological equipment have software components including CPS operational system from different manufacturers. Wide range of CPSs, mobile means of communication and integrated systems program applications practically exclude the creation of compact security system to provide high-level protection of the company resources.

In addition, equipment from different manufacturers normally functions properly with additional software components to automatize the production activity. When a heterogeneous cyber and physical production is projected, it necessary to choose inter-operable components to minimize the company necessity of additional software application.

4. Conclusion

The Industry 4.0 production company information security is a multi-factor task with deep automatizing of technological processes and application of cyber and physical systems to implement digital ways of automatic control. By implementing digital technologies in production, the task of company information security increases.

The most effective protection against the company information security threats are machine and program ways of the company information protection.

The machine ways of the company information protection are based on the attacker counter-measures technologies, which block potential channels of the information flow away (production data) and channels to implement malicious software into production infrastructure. The machine ways of information protection help to identify potentially susceptible objects of the company infrastructure, investigate methods efficiency and means of protection and also to detect attempts of attackers.

The program ways of the company information protection are based on the technologies of the functionality control and the operator restricted access rights with CPS to the production data and production components. The access rights restriction for the production company personnel is performed using an anti-virus software, decryption and cryptography software, virtual nets and inter-net screens software and other.

Unlike the machine information protection means dedicated to protection of the production infrastructure in general, the program means of protection can be locally placed as a part of the production objects with their own calculation means. Protection means localizing can increase the company security level with the ability to detect an attempt of unauthorized information access before the malicious software distribution in all production components.

An important aspect of the cyber and physical production information security increase is an operator and CPS actions protocols, which will be analyzed later by the artificial intelligence system placed in the company cloud environment. The analysis of production process participants' behavior scenarios may detect in time an attempt of the attacker's unauthorized access to the production information and operator and CPS actions correctness in production.

The practical realization of information security of a digital company having cloud services and applications distributed in a territory including a trans-country space is based on some regulations and standards, which exist in some countries only and in international affairs. In this regard, to provide the Industry 4.0 digital factory, smart factory and virtual factory information security it is necessary to develop the standards of new generation compliant with the condition of digital economy in the international level.

References

[1] Elvik R and Goel R 2019 *Accident analysis and prevention* **129** 136-147
[2] Lai X, Balakrishnan A, Lange T, Jenihhin M, Ghasempouri T, Raik J and Alexandrescu D 2019 *Microprocessors and microsystems* **71** 102867
[3] Bogatyrev V A, Parshutina S A, Poptcova N A and Bogatyrev A V 2016 *Communications in computer and information science* **678** 337-348
[4] Rauch E, Linder C and Dallasega P 2019 *Computers and industrial engineering* **2019** 127 105644
[5] Singh P and Singh L K 2019 *Nuclear engineering and technology* **51** 1289-96
[6] Ribeiro da Silva E H, Shinohara A C, Pinheiro de Lima E, Angelis J and Machado C G 2019 *Procedia CIRP* **81** 240-245
[7] Gurjanov A V, Zakoldaev D A, Shukalov A V and Zharinov I O 2018 *IOP Conference Series: Materials Science and Engineering* **327** 022110
[8] Zakoldaev D A, Shukalov A V, Zharinov I O and Zharinov O O 2018 *Journal of Physics: Conference Series* **1015** 052033
[9] Mogos M F, Eleftheriadis R J and Myklebust O 2019 *Procedia CIRP* **81** 624-629
[10] Li D, Landstrom A, Fast-Berglund A and Almstrom P 2019 *Procedia CIRP* **84** 380-386
[11] Alcacer V and Cruz-Machado V 2019 *Engineering science and technology, an international journal* **22**(3) 899-919
[12] Barreto L, Amaral A and Pereira T 2017 *Procedia manufacturing* **13** 1245-52
[13] Kravchenko P and Oleshchenko E 2017 *Transportation research procedia* **20** 367-372

AUTHOR INDEX

Ageev, E V ... 39, 45
Ageeva, E V ... 39, 45
Aleksandrova, N S ... 13
Altukhov, A Yu ... 39, 45
Arsentiev, G O ... 85
Arsentiev, O V ... 85
Arshinskiy, L V ... 90
Bataev, I A ... 13
Bataronov, I L ... 50
Bogdanovich, V I ... 20
Boldyrev, I S ... 71
Bonnard, E ... 151
Borodina, E ... 28
Buslovich, D G ... 1
Dezhin, V V ... 50
Dontsov, Yu V ... 1
Dunaev, M P ... 90
Dyussebayev, I M ... 145
Efremenkov, E A ... 145, 151, 157
Efremenkova, S K ... 145
Emurlaev, K I ... 55
Emurlaeva, Yu Yu ... 13
Galiev, I I ... 79
Galin, N E ... 157
Gerike, B L ... 109
Giorbelidze, M G ... 20
Gurjanov, A V ... 67
Ivanov, I V ... 55
Karmanova, A ... 7
Khomyakov, M ... 7
Klishin, V I ... 109
Kornienko, L A ... 1
Kozlov, V N ... 139
Kuznetsov, F Y ... 117
Lazurenko, D V ... 7
Liu, S ... 139
Lustenkov, M E ... 122
Lustenkova, E S ... 122
Lyalkina, G B ... 166
Makarychev, N S ... 166
Mokrushev, A A ... 109
Nikolaev, A V ... 166
Ogneva, T S ... 7
Panin, S V ... 1
Perevoshchikov, F P ... 128
Petrov, A ... 28
Petrov, G I ... 79
Prokofieva, E ... 28
Prudnikov, A P ... 133
Rudenya, E A ... 55
Ruktuev, A A ... 7
Shabo, A K ... 128
Shalamov, V G ... 71
Shibinskiy, K G ... 157
Shiler, A V ... 79
Shiler, V V ... 79
Shilin, A A ... 61, 128
Shukalov, A V ... 67, 98, 171
Smetanin, S D ... 71
Thoemmes, A ... 55
Tkachenko, E I ... 55
Vakulenko, S ... 28
Vuong, N V ... 128
Zaitsev, A V ... 103
Zakamaldin, A A ... 61
Zakoldaev, D A ... 67, 98, 171
Zhang, J ... 139
Zharinov, I O ... 67, 98, 171
Zharinov, O O ... 98, 171